AN EAR TO THE GROUND

UNDERSTANDING YOUR GARDEN

KEN THOMPSON

TRANSWORLD PUBLISHERS
61–63 Uxbridge Road, London W5 5SA
A Random House Group Company
www.transworldbooks.co.uk

First published in Great Britain
in 2003 by Eden Project Books
an imprint of Transworld Publishers

This paperback edition published in 2008

Illustrations by Mary Claire Smith
Design by Julia Lloyd

A CIP catalogue record for this book
is available from the British Library.

ISBN 9781903919200

Addresses for Random House Group Ltd companies outside the UK
can be found at: www.randomhouse.co.uk
The Random House Group Ltd Reg. No. 954009

The Random House Group Limited supports The Forest Stewardship
Council (FSC), the leading international forest-certification organization. All our
titles that are printed on Greenpeace-approved FSC-certified paper carry the FSC logo.
Our paper procurement policy can be found at:
www.rbooks.co.uk/environment

Typeset in 11.5/13 pt Granjon by Falcon Oast Graphic Art Ltd.
Printed and bound by CPI Group (UK) Ltd, Croydon, CR0 4YY

2 4 6 8 10 9 7 5 3 1

To two great Sheffield ecologists

Oliver Lathe Gilbert
7 September 1936–15 May 2005

Arthur John Willis
11 January 1922–20 June 2006

Contents

Preface

'Histories have previously been written with the object of exalting their authors. The object of this History is to console the reader. No other history does this.'

1066 AND ALL THAT

SUBSTITUTE 'GARDENING BOOKS' FOR 'HISTORY' AND THE first three sentences of the preface to Sellar and Yeatman's classic sum up my feelings about many gardening books. Comparing the carefully composed pictures of acres of manicured lawn and herbaceous borders at the peak of their July perfection with one's own humble plot, it's hard not to be discouraged.

The time therefore seemed ripe for an altogether different kind of gardening book, and this is it. Its aim is simple: to allow you to understand why your garden is the way it is. If your lawn is full of weeds, it will explain why this is an entirely natural state of affairs. If your attempt at wildlife gardening has produced nothing other than a patch of rank grass and thistles, it will show you why this outcome could have been avoided only by almost superhuman effort. Often it gently suggests how things might be improved, but this is not its primary aim.

It is perhaps worth mentioning at the outset that this book differs in another important respect from nearly all other gardening books. It takes its inspiration not from the gardening literature, and only rarely from the agricultural or horticultural literature. Most often it draws on botany, ecology and natural history in the broadest sense. In other words, from a literature that seeks not to control the natural world, but merely to understand it.

Armed with this volume, you will be able to appreciate your garden for what it really is – a large, unplanned and slightly out-of-control scientific experiment. You will then be properly equipped to rejoice at your successes and, more importantly, learn from your failures.

Gardens,
Gardeners
and Plants

A Nation of Gardeners

BY A WIDE MARGIN, GARDENING IS BRITAIN'S MOST POPULAR leisure activity. Gardens occupy a greater proportion of the land area in Britain than in any other country. One quarter of the land area of the average British city is private garden. Tourists visit Britain from all over the planet just to see our great gardens. The landscape garden is arguably the only truly original British contribution to the history of art. If gardening is ever admitted to the Olympics, there is no doubt who will win more than a few medals, without the assistance of any performance-enhancing substances beyond some compost and a mug of tea.

How have the British come to occupy this unique position? The explanation involves at least three separate elements. The first is our remarkably mild climate. Britain's position on the westerly edge of Eurasia, bathed by the warm Gulf Stream, means we can easily grow plants from all over the temperate and much of the subtropical world. Now and then some of our more adventurous plantings are killed by a hard winter (see 'Is it hardy?' page 127), but by and large this only adds extra interest to our perennial obsession with the weather.

The second part of the jigsaw is that, paradoxically,

despite having a climate where almost anything will grow, our native flora is remarkably impoverished. We have our recent climatic history to thank for this. Several times over the last few million years much of Britain has been scraped clean by glaciers. The parts that remained clear of ice were Arctic tundra. During these ice ages (which were far longer than the intervening warm periods) the sea level fell and Britain was joined to Europe, but as the ice melted we once more became an island, most recently just 10,000 years ago. Each time this happened many plants forced out by the ice did not manage to return. However, the glaciations also had more far-reaching effects. Each time the ice retreated, warmth-loving plants advanced northwards from their southern refuges. This advance was easy and rapid, because the land left by the ice was not densely occupied by trees. But the same was certainly not true when the ice returned, and the warmth-loving plants were generally unable to retreat, so most simply died where they stood. Those that survived the warm periods did it by moving up and down southern mountains, retreating to nearer the summits during the warm periods and returning to lower levels during the cold ones. As a result there were always some plants in the south, ready to move north when the opportunity arose. Not all Europe's original complement of plants managed to work this trick successfully during several successive glaciations, and many became extinct. One example familiar to gardeners is the genus *Magnolia*, once abundant in Europe (including Britain) but now extinct here. Another is *Rhododendron*, now with only four species in Europe (out of 850 worldwide), none of them much grown in gardens.

Many familiar garden trees and shrubs survive in Asia and North America but have disappeared from Europe, including *Sassafras*, *Lindera*, *Illicium*, *Mahonia*, *Stewartia*, *Hamamelis*, *Itea*, *Hydrangea*, *Physocarpus*, *Gleditsia*, *Pieris*, *Aralia* and *Liriodendron*.

A third element is Britain's long history of colonial expansion and exploration. British art galleries, museums and gardens are testament to the jackdaw-like acquisitiveness of the British abroad. British gardens have been enriched by countless plants brought back by waves of colonial expansion, from the Americas, Australia, New Zealand and the Himalayas. The *RHS Plant Finder* now lists over 80,000 species and cultivars for sale in British nurseries and garden centres. Edinburgh Botanic Garden alone grows over 17,000 species (ten times the number in our native flora), Kew substantially more than that. Nor is the process complete. Every year new species are discovered and many, sooner or later, find their way into British gardens. A dramatic recent example is the Wollemi pine (*Wollemia nobilis*). This remarkable relative of the monkey puzzle was discovered in 1994 by David Noble, from the NSW National Parks and Wildlife Service, while walking in the Wollemi National Park, about 150 kilometres north of Sydney. Despite being one of the world's rarest plants – down to fewer than one hundred adult trees at the time of its discovery – it looks set to follow in the footsteps of that other great survivor, *Ginkgo biloba*, by enjoying a completely new lease of life in cultivation. You can buy one for your garden, but at the time of writing it will cost you £97, so my advice is to be satisfied with the one at Kew for the time being.

A BRIEF HISTORY OF PLANTS

Plants invaded the land about 450 million years ago. For the following 100 million years, an almost unimaginably long period, one group of primitive plants succeeded another. Much of this evolution paralleled what was going on in the animal kingdom, and involved an increasing independence from water. In mosses, ferns and their relatives, part of the life cycle needs free water. These are the amphibians of the plant world. The big breakthrough, which happened about 350 million years ago, was the evolution of seed plants, in which male gametes (pollen) are transferred directly from one plant to another, without having to swim.

Seed plants were a big success, and soon developed into many different forms. Many of these early seed plants are now extinct, and of those that remain only three groups are of much interest to gardeners. Conifers are by far the most successful primitive seed plants, and are also the most widely cultivated in gardens. Cycads are of interest only to gardeners in the tropics and subtropics, or those with large conservatories. One isolated survivor, not closely related to any other plant, is *Ginkgo biloba*, the maidenhair tree. Fossils indistinguishable from *Ginkgo* are known from all over the world, including Europe, but it survived until modern times only in China. In fact, even there it has long been extinct in the wild and survives only in cultivation. A striking columnar tree with unusual leaves and a beautiful yellow autumn colour, *Ginkgo* is worth growing in any but the smallest garden. But beware – you will find that many books warn about the smell of *Ginkgo* fruits, and they are right. Like many primitive seed plants, *Ginkgo* has separate male and female plants, and if you plant a female along with a male you will live to regret it – trust me. Another curious feature of *Ginkgo* is that it's arguably the world's most frequently misspelled plant name, at least if the internet is any guide. Googling *Ginkgo* produces about 5 million hits, but *Gingko* produces a respectable 1.7 million, so it looks like there are lot of confused *Ginkgo* fanciers out there. *Fuchsia* isn't far behind though, with 6.2 million hits for *Fuchsia* and 1.6 million for *Fuschia*.

To the would-be Jurassic Park gardener, however, all these primitive plants had one overwhelming defect: they had no flowers. For 310 million years, the world was just plain green. Mosses, ferns and other spore-producing plants don't need pollinating (they don't have pollen), and all primitive seed plants were (and are) wind-pollinated. But around 140 million years ago, near the beginning of the Cretaceous period, one group of plants began a partnership with insects that was eventually to make modern gardening possible. We don't know exactly what these plants looked like, nor do we know exactly what happened, but we can make an educated guess. Primitive insects, almost certainly beetles, began to visit primitive reproductive structures (we can't yet call them flowers) to eat pollen and ovules (immature seeds). Both of these are rich in protein and are a much better food source than leaves. In wandering between 'flowers' and between plants, these beetles were the first insect pollinators. Natural selection quickly favoured plants that attracted pollinators, with colour and scent, and gave them something to eat other than the flower itself, i.e. nectar. These new kinds of plants are now called angiosperms, or simply flowering plants, and there are many more different kinds than all the more primitive plants put together.

The partnership with insects was a runaway success, but after a period of rapid evolution things slowed down a bit. Between about 75 and 50 million years ago, however, the evolutionary tempo picked up speed again. The reason? The evolution, during this period, of the great groups of specialist insect pollinators: moths, butterflies and modern flies and bees. Flowering plants and insect pollinators are one of evolution's great double acts, and one for which gardeners have every reason to be profoundly grateful. As a postscript, it is worth noting that whatever finished off the dinosaurs about 65 million years ago had little obvious impact on the rise of the flowering plants.

It may seem obvious that flowers look, and often smell, attractive. After all, natural selection has made them that way. But wait a minute. Attractive to whom? Flowering plant evolution was essentially complete at least 45 million years before humans appeared. It was not at all inevitable that things that look and smell good to insects should also look and smell good to people. Forget for a moment the senses of sight and smell, and consider hearing. Could you even begin to write a symphony that would appeal to a bee? Do more flies turn up in your living-room when Beethoven is on the stereo? Or Burt Bacharach (maybe insects go for easy listening)? The truth is that insect brains and senses are quite alien, and it is astounding that flowers appeal equally to insects and humans. Consider how things might have turned out had some groups of flies become the main insect pollinators. Flowers pollinated by insects that lay eggs in carrion or dung have evolved to look and smell like carrion or dung. Many a young cactus enthusiast has had his entire collection thrown out after his *Stapelia* made the mistake of flowering. *Stapelia*s are African or Asian cactus-like plants with flowers that have to be smelt to be believed, and can easily make an entire house uninhabitable.

THE GHOST OF MOAS PAST

Many of the more outré corners of the plant kingdom are in the tropics, and thus unavailable to temperate gardeners or at least those without enormous greenhouses. We can't grow strangling figs, fearsome-spined rattans or the horribly smelly metre-wide flowers of *Rafflesia*.

On the other hand, we can grow a few interesting oddities, like *Corokia cotoneaster*. If you grow this, you'll know it's a medium-sized evergreen shrub with small, starry yellow flowers and red berries. Pretty, rather than dramatic. You'll also know that its real claim to fame is betrayed by its common name, the wire-netting bush. As the name suggests, its growth habit is hard to describe yet instantly recognizable: it looks like a tangle of wire netting. The technical reason for this is that the slender, wiry branches grow at an unusually acute angle, often at ninety degrees to the main stem. The side branches are also unusually long in relation to the main stem. The botanical jargon for this phenomenon is 'divaricating'. *Corokia* comes from New Zealand, where divaricating plants are curiously common and make up almost 10 per cent of the native woody plants, a higher proportion than anywhere else in the world.

The obvious question is: why? What is it about New Zealand that has favoured the evolution of such a strange growth form? It's been suggested that it's something to do with climate, or soils, or just an accident, but none of these explanations stands up to close scrutiny. It's got nothing to do with geography, since divaricating plants are absent from other Pacific islands. The most persuasive theory is that browsing animals are to blame. New Zealand completely lacks native browsing mammals such as horses or deer. The main browsers (now extinct) were the moas, large relatives of emus, cassowaries, rheas and ostriches.

The divaricating habit has the effect of creating an almost impenetrable tangle of twigs, with most of the leaves tucked away inside this tangle. This habit may have been particularly effective against browsers like moas that couldn't bite and chew their food like mammals, and instead relied on pulling bits off plants. Of course, since we can't observe moas in action, we will never know for sure if this theory is correct, but there are several other pieces of supporting evidence. First, spiny shrubs are unusually scarce in New Zealand, and spines are not much use against an animal with a long horny bill. Second, several divaricating plants have non-divaricating varieties on offshore islands that lacked moas. Third, there are nine species of divaricating trees, all of which are divaricating only as juveniles, and change to a non-divaricating habit when they get to be about 3–4 metres tall (about the height needed to escape the taller moas). Finally, the only other place with a concentration of wire-netting plants is Madagascar, which had its own giant ostrich relatives, the extinct elephant birds.

GARDENERS' TIP

Of all the New Zealand divaricating shrubs, Corokia cotoneaster *is the one most widely available to gardeners. Others that are in cultivation but harder to find are* Coprosma acerosa, C. propinqua *and* Discaria toumatou. *All three are grown for their growth habit, and none is very interesting in flower.* Discaria *is almost worth growing for its common name alone – wild Irishman.* Sophora microphylla *is a small tree with a divaricating juvenile form and yellow flowers in spring. It would make an unusual alternative to the ubiquitous laburnum.*

WHAT'S IN A NAME?

There is an interesting tension between gardeners and botanists on the subject of plant names. Essentially this is because the two groups use names for different purposes. Gardeners are much more interested in the original purpose of plant names, which was to give plants universal, unambiguous labels that could be used and understood by anyone. Scientific names are in Latin because when the present system was developed Latin was the universal scientific language. In fact, even the name of the man who invented the system, Carl von Linné, is usually Latinized to Linnaeus. Recently, it has become much more important to botanists that plant names accurately reflect the evolutionary relationships between different species. More of that later; but first, what do you need to know to understand plant names?

The science of plant classification is called *taxonomy*. Plant names are hierarchical, with different levels fitting inside each other like Russian dolls. Botanists recognize a plethora of levels, but only three are important to gardeners. These are **family, genus** and **species.** Family names are written in normal type, but genus and species names are always italicized. Don't ask me why; they just are. Classification of plants is based largely on flowers, and so families are natural groupings of plants with essentially similar flowers. For example, the Cruciferae (now Brassicaceae), or cabbage family, contains weeds (shepherd's purse, hairy bittercress), vegetables (cabbage, radish) and garden plants (wallflowers) with four petals and usually long seed pods. The Leguminosae (now Fabaceae), or pea family, comprises all the plants with pea-like flowers. Knowing the family a plant is in can tell you quite a lot, but exactly how much varies from family to family. The cabbage family, for example, is entirely herbaceous, with a distinct tendency to be weedy. The pea family, on the other hand, is much more variable, containing everything from annuals (sweet pea), herbaceous perennials and vegetables (lupins, runner beans) through shrubs (gorse) to trees (laburnum).

The genus is perhaps the most fundamental level of

plant classification, and at the same time the word itself is hardly known to most gardeners, even though generic names (*Geranium*, *Rhododendron*, *Delphinium*) are the staple of any gardeners' conversation or television programme. To a large extent, this is because the word 'genus' has been usurped in popular parlance by the word 'family'. Thus when gardeners talk about the daffodil family, they mean, well, daffodils (*Narcissus* species). When botanists talk about the daffodil family, they mean Amaryllidaceae, including (among others) *Narcissus*, *Galanthus*, *Leucojum*, *Amaryllis* and *Nerine*.

Before leaving families and genera, it is worth noting that gardeners often use names that have no formal taxonomic status, yet are nevertheless extremely useful, and are normally collections of more-or-less related genera, e.g. brooms (including *Cytisus*, *Genista* and *Spartium*). Some of these groups are so useful that gardeners even write books about them and form societies dedicated to their cultivation, e.g. heathers (*Calluna*, *Erica*, *Daboecia* and many others). Other names have no connection with taxonomy at all. More than a dozen different plants, in nearly as many families, have a claim to be called laurels.

Within the genus lies the species. Thus within the genus *Viburnum* are several species well known to gardeners, including *Viburnum tinus*, *Viburnum davidii* and *Viburnum opulus*. In a list like this, or anywhere the genus is understood, it is usually shortened to its initial, e.g. *V. davidii*. Garden plants are often not just the plain species. They may be selected cultivars (a contraction of 'cultivated variety'), e.g. *V. tinus* 'Eve Price'. Notice that the cultivar name is not

Latin; in fact the taxonomic rules no longer allow Latin names for cultivars, although any other language will do, e.g. *Prunus* 'Tai Haku'. Often a cultivar's history is so complex that it can now only be referred to a genus rather than a species, e.g. *Rosa* 'Fragrant Delight'. A plant can have only one 'official' cultivar name, but plant breeders can register other 'trade' or 'fancy' names, sometimes several in different countries, especially if the original cultivar name is in a foreign language. Sometimes the trade name is simply a direct translation of the cultivar name; for example the German *Bergenia* cultivars 'Silberlicht' and 'Morgenröte' have the trade names **Silverlight** and **Morning Red** in the UK. In an attempt to avoid confusion between cultivar and trade names, the former are enclosed in quotation marks and the latter are usually written in a distinctive **sans serif** typeface. Sometimes, confusingly, a plant has two or even more cultivar names, although technically only one is correct. The excellent penstemon 'Garnet' made headway in the UK only after it was given its English name in 1950, but 'Andenken an Friedrich Hahn' remains its correct name.

Garden plants may also be hybrids of two species, in which case the name may be just the two specific names separated by '×'. More often hybrids are given a new specific name, prefixed by '×' to show it is a hybrid. Thus the hybrid of *Viburnum farreri* and *Viburnum grandiflorum* is called *V.* × *bodnantense* (named from Bodnant garden in Wales, where the cross was first made in 1934). Hybrids may also have cultivars, e.g. *V.* × *bodnantense* 'Dawn'. Hybrids of two genera also occur, and here the '×' goes in front of the new genus name. Relatively few such 'bigeneric' hybrids are grown in gardens, but one very attractive example is × *Fatshedera lizei,* the hybrid of *Fatsia japonica* and ordinary ivy, *Hedera helix*.

Only quite closely related genera form bigeneric hybrids. Some groups of plants rarely do, especially those in which the genera are very ancient, and have therefore had plenty of time to become very different from each other. One such group is the conifers, which makes perhaps the most famous bigeneric hybrid of all, × *Cupressocyparis leylandii* (Leyland cypress), particularly surprising. Leyland cypress is a hybrid of Nootka cypress, *Chamaecyparis nootkatensis*, and Monterey cypress, *Cupressus macrocarpa*. Recent research suggests that the former has been misclassified, although there is still some doubt about exactly where it belongs. If, as some taxonomists suspect, it really belongs in the genus *Cupressus*, the conifers will lose their only bigeneric hybrid, and Leyland cypress will become *Cupressus* × *leylandii*, which at least will make it easier to spell. Whatever its origins, however, Leyland cypress is still a remarkable example of hybrid vigour, which sometimes makes hybrids much bigger and better than their parents. Just how much bigger is anybody's guess, since the oldest Leyland cypresses in existence are still growing and their final height is unknown.

GARDENERS' TIP

If you want, say, a Shasta daisy, and you don't care if it's large or small, double or single, with petals entire or frilly, then that's all you need to know. If you want a particular Shasta daisy, such as the creamy-yellow 'Sonnenschein', then you'll just have to get to grips with plant names.

You'll also have to make sure you deal with nurseries who know and grow their plants. You won't, except by extraordinary good fortune, find what you want in your local garden centre. To find a particular plant, try The Plant Finder, *published by the Royal Horticultural Society and also available on their website (www.rhs.org.uk).*

WHY FLOWERS?

In the previous section I happened to mention, more or less in passing, that plant classification is based largely on flowers. But why? Why can't plants be classified on the basis of, say, their fruits or leaves? To some gardeners, the emphasis on a few relatively obscure flower parts is just a botanical conspiracy designed to make everyone else look stupid. After all, if you put a buttercup, a delphinium and a clematis side by side, it's hardly obvious that they are members of the same family. In fact, of course, plants can be classified in almost any way you like, and they frequently are. My *RHS Gardeners' Encyclopedia of Plants and Flowers* arranges plants by size, flower colour and time of flowering. Many other books arrange plants by their preferred habitat (plants for shade, boggy places, etc.). All these classifications are perfectly acceptable, but they are all *artificial*. To see what I mean, let's look at the most artificial classification of all. The *RHS Plant Finder* arranges its contents alphabetically. This arrangement allows you to find the plant you want as quickly as possible, and of course requires no index. But it tells you nothing at all about a plant, beyond the letter its name begins with. *Daphne* is next to *Davidia*, and rhododendrons rub shoulders with sumachs (*Rhus*).

I suppose the opposite of an *artificial* classification should be a *real* one, but in fact botanists talk about *natural* classifications. A natural classification tries to use *all* the available information, and aims to group together plants that resemble each other in as many ways as possible. In much the same way, if you were asked to arrange a roomful of people into similar groups, you wouldn't do it only by earlobe shape or shoe size. Natural classifications are therefore *predictive*; that is, the classification tells you many unspoken things about the plant in question. Knowing whether a plant is a brassica or a *Buddleja* tells you nearly all you need to know about it, in a way that knowing its name begins with 'B' does not. Now, it just happens that centuries of trial and error have convinced botanists that, *in general*, floral characteristics are the most predictive. I should say *some* floral characteristics, since the exact details of, say, petal size, colour and shape are much influenced by pollinator behaviour. This variation is often very striking (remember the buttercup and the delphinium), and may at first sight completely conceal the underlying kinship.

Of course, botanists can be fooled like anyone else, and an apparently fundamental similarity is not always all it seems. The sacred lotus, *Nelumbo*, has always been lumped with the water-lilies because, well, it just looks like a water-lily. However, new molecular classifications reveal that

Nelumbo and the water-lilies are not even distantly related. In fact *Nelumbo* turns out not to have any really close relatives at all, and its nearest relations are all trees and shrubs in the plane (*Platanus*) and *Protea* (*Embothrium*, *Grevillea*) families. The remarkable similarity of *Nelumbo* and the water-lilies is simply an excellent example of convergent evolution, since both exploit the aquatic habitat in much the same way.

A ROSE BY ANY OTHER NAME

Gardeners frequently complain that botanists keep changing the names of plants. While it is tempting to think they do this to keep themselves in work, or just out of mischief, there is no truth in the latter charge and only a little in the former.

Name-changes have a number of origins. One is the 'rule of priority'. Modern plant naming dates from the publication in 1753 of Linnaeus's book *Species Plantarum*. For plant names to exist at all they must be published in a recognized scientific journal, and the correct name for a plant is the *first* name published after 1753. So one reason names change is that botanical detectives keep unearthing earlier, previously overlooked, names. Fortunately, in a rare outbreak of common sense, the rules allow existing names to be kept if replacing them would cause too much chaos. So even though some bright spark discovered that an earlier (and therefore correct) name for *Freesia* was *Anomatheca*, it was decided to keep the well-known name, rather than risk simply being ignored by florists everywhere.

A more serious problem arises from the different purposes of botanists and gardeners. For reasons that we don't need to go into, botanists are increasingly concerned that the classification of plants should reflect evolutionary relationships. In other words, species within genera should be more closely related to each other than to those in other genera, and so on up the taxonomic scale. This requirement transcends all others, and if it means upsetting the existing time-honoured names, then, well, it's the names that have to change.

In the past plant classification has been based on morphology – the study of forms: number of petals, ovary position, and so on. Inevitably there has been scope for disagreement about which plant characters are most important, which has allowed botanists to revise existing species, genera or whole families and give them new names. Acceptance of such changes has proceeded by consensus, and changes that have not met with widespread acceptance have been ignored. Recently, however, molecular taxonomy has arrived. We can now look directly at proteins and DNA and classify plants with much greater certainty than in the past. The unpleasant message for gardeners is that this is going to upset a lot of apple-carts. For example, maples and horse chestnuts, both proud owners of their own families when I was younger, are now not only in the same family as each other but also a bunch of tropical fruit trees, including lychee and rambutan. A consistent change revealed by the DNA is that lots of plants are not as closely related to each other as we used to think, which means lots more families; one standard reference work that used to contain 306 families now contains 506. As a result, some families that used to be vast, sprawling empires are now reduced to mere principalities. The lily family (Liliaceae), made up of close to 300 genera at it peak, is now down to a meagre sixteen, with lily of the valley, bluebells, hostas, day lilies, agapanthus and alliums (among many others) now all given families of their own. Mind you, the lily family still includes lilies, tulips and fritillaries, so I suppose it can't really complain.

GARDENERS' TIP

Try to be philosophical about botanists messing about with plant names. It's going to get a lot worse before it gets any better. Looking on the bright side, it will be years, maybe decades, before the latest changes work their way through to gardening books and garden centres.

LATIN FOR BEGINNERS

As Peter Cook once famously said, he could have been a judge, but he didn't have the Latin. Many gardeners feel the same. A basic knowledge of botanical Latin, however, should be the aim of every keen gardener. You will have a better idea of what you are growing, and Latin names (when correctly decoded) often tell you something interesting about the plant to which they are attached. Having said that, many genus names are just, well, names. Like John or Susan, they don't really mean anything – they are just what the plant has always been called, often since at least Roman times. *Salvia* and *Rosa*, for example, are what the Romans called sage and various roses. Many generic names commemorate botanists or politicians, for example *Fuchsia* (Leonard Fuchs, a sixteenth-century German botanist – remembering this, by the way, will help you to spell it correctly), *Grevillea* (C. F. Greville, one of the founders of the Royal Horticultural Society), *Gunnera* (J. E. Gunner, a Swedish botanist) and *Tulbaghia* (Ryk Tulbagh, governor of the Cape of Good Hope). Some generic names mean something, but it isn't always obvious what; for example *Aquilegia* (from *aquila*, Latin for eagle; the spurred petals are supposed to resemble eagle talons) and *Gypsophila* (chalk-lover, from a preference for chalky soils).

Species names are much more often descriptive, if unhelpful. To take some very common examples: *officinalis* (of apothecaries' shops, hence herbal or medicinal), *pratense* (of fields or meadows), *sylvatica* (of woods or forests) and *paludosa* (of bogs or marshes). Some are straightforward descriptions, such as *rotundifolia* (round leaves), *octopetala* (eight petals), *nigra* (black), *muralis* (on walls), *mirabilis* (wonderful) and *lunatus* (half-moon shaped). Colours are sometimes not at all obvious – *caerulea* is blue, *flava* and *lutea* are both yellow – and sometimes confusing: *nivalis* can mean either snow-white or growing in snow, and sometimes both. Many names refer to the country or region of origin, either obviously (*japonica*, *chilense*, *mediterranea*, *missouriensis*) or obscurely (*nootkatensis*, from Nootka Sound, British Columbia; *thuringiaca*, from Thuringia (mid-Germany); *tucumaniensis*, from Argentina). But don't accept all such names at face value – the authors of names weren't always too sure where the plant came from. Sarnia is an old name for Guernsey, so naturally the Guernsey lily is *Nerine sarniensis*, which would be fine if the plant in question weren't a native of South Africa. There are many examples of this sort: Aleppo pine doesn't come from Aleppo, and Deptford pink never grew in Deptford.

Three positively useful species names are *sativa* (cultivated), *edulis* (edible) and *esculentum* (good to eat). Thus we have *Boletus edulis*, arguably one of the best-tasting toadstools, and *Lycopersicon esculentum* (tomato). Most *sativa* species are edible too (e.g. *Pisum sativum*, pea, and *Lactuca sativa*, lettuce), but some have other uses (e.g. *Cannabis sativa*). Your opinion of *sativa* as a guide to useful plants may be coloured by your attitude to parsnips (*Pastinaca sativa*). The most overworked species name is *vulgaris* (common). Botanists are compelled to use this because of a quirk of the rules for naming plants, which do not allow the genus and species name to be the same. Zoologists don't have this problem, so we have *Buteo buteo* (buzzard), *Meles meles* (badger) and *Troglodytes troglodytes* (wren). If the primrose had been an animal, it would doubtless have been *Primula primula*, but being a plant it's stuck with *Primula vulgaris*. Which reminds me of my favourite animal trivia question: which bird has the Latin name *Puffinus puffinus*? Answer: the Manx shearwater, which just goes to show you shouldn't take all Latin names at face value.

Biological jargon is littered with such traps for the unwary. What, for example, is batology? You'll not be surprised to hear it is nothing to do with bats. It is, in fact, the study of brambles. Why, you might reasonably ask, do we need such a word? After all, as far as I know there's no word for the study of violets, say, or buttercups. The reason is brambles' peculiar sex life. Although bramble flowers are attractive to many insects, they reproduce by apomixis, which means the ovules are rarely fertilized and seedlings are usually exact copies of their parent (dandelions are the same). The result is hundreds of 'microspecies', their names known only to a few dedicated (or bonkers, depending on your point of view) batologists. One practical consequence, known to all keen blackberry pickers, is that brambles are extremely variable and some microspecies are delicious while others are rubbish.

Species names, like genera, often commemorate famous people. Botanists have tended to be quite restrained about this, sticking mainly to naturalists or explorers. For example *Berberis darwinii*, *Pieris forrestii*, *Rhododendron wardii* and *Acer davidii*, after Charles Darwin and the plant hunters George Forrest, Frank Kingdon Ward and l'Abbé Armand David respectively. All first-class garden plants, by the way. Zoologists on the other hand, admittedly with the excuse of far more species to name, have tended to let themselves go. So we have *Bufonaria borisbeckeri* (a sea snail), *Calponia harrisonfordi* (a spider), and *Baeturia laureli* and *B. hardyi* (cicadas). The Canadian palaeontologist Greg Edgecombe and his colleagues are responsible for four fossil trilobites named after the rock band the Ramones, another five named after the Sex Pistols, and *three* named after the Beatles. I don't know whether George Harrison had done something to offend them, or whether they were just one trilobite short.

Finally, don't expect nurseries and seedsmen to pay any attention to Latin names if they think they might get in the way of selling you something. In a catalogue that landed on my doormat recently, *Abeliophyllum distichum*, a creamy-white, scented relative of *Forsythia*, had become the 'rare white *Forsythia*'. At least the real name was still mentioned in the small print, but on another page *Nectaroscordum siculum*, an exceptionally handsome and distinguished relative of the onion, had been transformed completely into 'Mediterranean Bells'. Its real name was nowhere to be seen, which is a pity if you want to look it up, because the name Mediterranean Bells seems to be one the nursery just made up. Another catalogue calls the same plant 'Honey Lily', and as far as I can see they just made that up too.

GARDENERS' TIP

Do not be afraid of Latin. It is the specialized jargon of gardening, and part of the fun of any hobby is getting to grips with the jargon, if only so you can show off to those who know less. Do not be put off by difficult pronunciation – nobody *knows how Latin should be pronounced except the Romans, and they aren't around to argue. And if you want to buy a* Paeonia mlokosewitschii, *just point.*

A FAMILY LIKENESS

There is no end to the variety of ways plants are arranged in gardening books. Apart from simple alphabetical order, favourites include size, flower colour and flowering season, while preferred sites (plants for shade, dry places, etc.) are also popular. Only rarely are plants arranged according to how they are related to each other, and in general, gardeners are not encouraged to think about family relationships of garden plants.

Botanists think quite differently. On meeting a plant for the first time, the first thing a botanist wants to know is 'What family is it in?' Once the family is known, everything falls into place, and there is nothing more disorienting than meeting a plant from an unfamiliar family. One ecologist claims to hold the world record for the largest number of different plant families consumed at one meal, but then maybe he doesn't have many competitors for that particular honour.

If you ever have the misfortune to be shipwrecked with a lifeboat full of botanists, here is a game you can play while you are waiting to be rescued. If you had to construct a garden entirely from a single plant family, which family would you choose, and why? This is such a difficult question to answer because almost everything runs in families. For example, if we assume that any garden should contain at least shrubs, climbers, herbaceous perennials and annuals, how many families can provide these four basic ingredients? The answer is not very many, and if we further stipulate that we want a decent selection of plants in each category, we are left with only one, the Leguminosae or pea family. Some other families might claim to have all four, but in every case they make a very poor job of one of them. For example, the daisy family (Compositae) gives you only one rather difficult climber (*Mutisia*), while the potato family (Solanaceae) is OK as long as you're happy with *Physalis* as your only herbaceous perennial. Moreover, these two families illustrate a general problem, which is that most families are essentially either woody or herbaceous. There are shrubs in the daisy and potato families, but compared to *real* woody families like maples or magnolias, you can tell they aren't really trying. Very few families (probably only the rose and pea families, and grasses if you count bamboos as woody) contain a wide range of woody *and* herbaceous garden plants.

Because things important to gardeners run in families, thinking botanically helps you to find your way around. Complete ignorance of plant family relationships is like going for a walk without a map, and leads you into errors akin to looking for Catherine Cookson in the non-fiction section of the library. Knowing what you like allows you to look for close or not-so-close relatives. Some of these will be similar to the plant you started with, and you will probably like them too. Others will be very different, and you may like them even more, or not at all. Either way, it's an adventure you wouldn't have had if you had stocked your garden entirely with the 'plant of the month' from the local garden centre. For example, if you like blue flowers, it is useful to know the very few families where shrubs with blue flowers are to be found. Blue-flowered herbaceous plants are spread across a much wider range of families, but even here, if you were particularly in earnest about blue, you would concentrate your energies on the few families that take blue really seriously (Campanulaceae, Boraginaceae, Gentianaceae). It also helps to know when families will be no help. For example, flower scent is not much of a family characteristic and can turn up almost anywhere, although obviously not in exclusively wind-pollinated families like grasses and many tree families.

By way of an introduction to plant families, the list below mentions some families in which the members that are not widely grown are at least as deserving as the ones that are. The list of families could have been much longer, and the list of plants that should be more widely grown could have been *very* much longer. Look on the list as a starter kit – the fun of this game is not in what I think you should grow, but in playing the game yourself.

IF YOU LIKE... IVY,

then try... other members of the Araliaceae.

If ivy sometimes looks a bit out of place in Britain, that's not too surprising. It is the only British member of a family that is otherwise mostly tropical. Many have bold, exotic-looking foliage, and *Aralia elata* would add style to any garden.

IF YOU LIKE ... WEIGELA, HONEYSUCKLE OR VIBURNUM,

then try... other members of the Caprifoliaceae.

If ever a family suffered from an embarrassment of riches, it is the Caprifoliaceae. Try the shrubby relatives of the familiar climbing honeysuckle. There are many fragrant winter-flowering shrubs, but none more delightful than *Lonicera fragrantissima*. Scent is a personal thing, but for me this is the most delightful smell you can grow. It's rather big and not much to look at in the summer, but no one with a large garden should be without it. Also, forget the rather untidy native elder and try some of the more ornamental species and cultivars. For a real conversation piece, try the blue elder (*Sambucus caerulea*), with beautiful blue berries.

IF YOU LIKE ... RHODODENDRONS,

then try... other members of the Ericaceae.

There are almost too many good plants in the huge heather family, but the calico bush (*Kalmia latifolia*) is one of the best, and grown far too rarely.

IF YOU LIKE ... LABURNUM,

then try ... other members of the Leguminosae.

The pea family is huge, indeed one of the biggest in the world, with trees, shrubs and herbaceous plants. Many of the woody members are a bit tender, or a bit untidy, but for a handsome small tree, and a change from the ubiquitous laburnum, try the Judas tree (*Cercis siliquastrum*).

IF YOU LIKE . . . FORSYTHIA,

then try . . . other members of the Oleaceae.

Fly over any British town in early spring and what do you see? A sea of yellow *Forsythia* × *intermedia* looking back at you from every suburban garden. Lilac, in the same family, is nearly as popular. A better plant, however, especially for those of you who don't enjoy pruning, is the slow-growing, evergreen *Osmanthus delavayi*. Unlike lilac and forsythia, it's attractive even when not in flower, and the white flowers are deliciously scented. Another small shrub, looking even more like forsythia, but with white or slightly pink scented flowers, is *Abeliophyllum distichum*. Widely available and easy to grow, it remains a mystery why it has never caught on with gardeners. Hmm . . . maybe it's that horrible name.

IF YOU LIKE . . . CLEMATIS OR AQUILEGIA,

then try . . . other members of the Ranunculaceae.

For sheer variety, the buttercup family is hard to beat. If only it contained a few shrubs, it would be the obvious candidate for the single family from which to create an entire garden. If you like the usual climbing clematis, then why not try the herbaceous (or sub-shrubby) varieties?

There are several species, all hardy, suitable for most soils and with the typical clematis flowers and fluffy seed heads. Also try the large meadow rues, such as *Thalictrum aquilegifolium*. As the name suggests, the leaves are indistinguishable from an aquilegia, but the flowers are quite different. Managing to be impressive, stately and at the same time delicate, meadow rue would be a valuable addition to any herbaceous border.

IF YOU LIKE ... NASTURTIUMS,

then try ... other members of the Tropaeolaceae.

Not easy to please, but worth the trouble, is the perennial *Tropaeolum speciosum*. It is sometimes stubbornly hard to establish, but if it likes you, you will be rewarded with a cascade of scarlet flowers draped over everything in sight.

LET THE PLANTS DO THE TALKING

'Right plant, right place' is a common phrase in gardening books. Not surprisingly either, for different plants have their own distinct preferences, for sun or shade, dry or damp, high or low pH. Yet many gardeners (including those who ought to know better) persist in trying to grow the wrong plant in the wrong place. My own garden is rather dry, or at least there's nowhere in it where plants of damp soil are likely to feel really comfortable. But does that stop me trying to grow them? No, it doesn't. Years ago, I took a fancy to *Camassia leichtlinii* and *Primula florindae*, the giant Himalayan cowslip. Both great plants – for moist soil, or by water. I put them both in the dampest spot in my garden, but all the while I knew it wouldn't be enough, and so it proved. I had to watch both give up flowering, then slowly fade away and die. At least the primula didn't cost me anything, since I grew it from free seed from a local gardening club.

Did that experience teach me a lesson? Well, yes and no. I have at least realized there's no point even trying to grow *Meconopsis* (apart from *M. cambrica*, which cannot be prevented from growing). On the other hand, last year I couldn't resist trying *Monarda didyma*, another lovely plant for a damp site. I planted two, and as much as I would like to tell you that just one of them survived, I cannot tell a lie. As for my bone-headed persistence in attempting to grow *Tropaeolum speciosum*, the less said the better.

All of which makes me particularly grateful for plants that are not only at home in my garden, but show it by abundantly self-seeding. Among the more disciplined type of gardener, self-seeders have got something of a bad name, but I've no time for such control-freakery. The beauty of self-seeders is that they will, given time, end up in the part of the garden they like best, without any effort on your part. Over the years, honesty, *Anemone blanda*, *Lathyrus vernus*, *Cyclamen hederifolium*, *Stipa tenuissima*, *Viola labradorica*, *Mentha requienii*, countless geraniums and a host of other plants have slowly rambled around my garden, seeking out their preferred spot. What's more, it's remarkable how often this process throws up plant combinations that look just right. There's no scientific reason for this, except that perhaps plants that *are* happy together also *look* happy together. Probably the most felicitous seasonal display in my garden is a spring mixture of deciduous azaleas, honesty, bluebells, forget-me-nots, red campion and wallflowers. Only the azaleas were planted by me, at least in that particular place. Even more surprising is the *Euphorbia amygdaloides* that turned up entirely of its own accord in the driest, shadiest corner of the garden. It's one of the few plants that *could* grow there, and it looks just right.

GARDENERS' TIP

Do not scorn plants that self-seed. Let them get on with it, and just pull them up and add them to the compost heap if they get out of control. But do approach self-seeders with your eyes wide open. Once you plant Meconopsis cambrica *or* Alchemilla mollis, *you've got them for life whether you like it or not, so do think carefully whether you really like them.*

Nor should you turn up your nose at 'common', popular plants. There is a good reason they're common: they have proved their ability to succeed under almost all conditions and, what's more, mass production means they're cheap. There's a list of twenty-four of the best on page 6 of D. G. Hessayon's Tree and Shrub Expert. *If you grow nothing else, your garden will be colourful and trouble-free.*

If you do want to grow more choosy plants, pick them to suit your conditions. Trying to alter the conditions to accommodate your favourite plants probably won't work, and will be expensive and hard work even if it does.

CHEMICAL COUSINS

Features you can see, like flowers, are not the only things that run in families. In some families, nearly all members are crammed with interesting chemicals, while other families are comparatively dull. No one is sure what these chemicals are for, but many of them are certainly there to deter herbivores, sometimes by being quite dangerously poisonous. Many poisonous plants are sources of medicinal compounds, and indeed the difference between a life-saving drug and a deadly poison is often just a matter of dose.

At one end of the chemical spectrum are families with no really toxic members at all, like the grasses, so it's OK to carry on chewing that grass stalk. Grasses *are* well-defended against animals, but their weapon of choice is the highly abrasive silica (the oxide of silicon and, in the form of quartz, the main constituent of sand). Eating grass is literally like eating sandpaper, and animals that eat grass must have specially adapted teeth. Bamboos can be up to 6 per cent silica, which is what makes bamboo canes so tough, although other grasses can contain even more. In fact bamboos suck so much silica out of the soil that it pays to replace it by regular mulching with grass. Soil-less composts, which contain little or no silica, are no good for bamboos.

GARDENERS' TIP

Before we leave grasses, a word about the recent fashion for grasses as garden plants. Many popular species are from warmer *climates, but a number are British natives, including quaking grass (*Briza media*), tufted hair grass (*Deschampsia cespitosa*), wavy hair grass (*D. flexuosa*), purple moor grass (*Molinia caerulea*) and meadow oat grass (*Helictotrichon pratense*). For a woodland setting, try wood millet (*Milium effusum*), nodding melick (*Melica nutans*) or wood melick (*Melica uniflora*). Although the grasses sold in garden centres are often related species or cultivars, they don't look much different from the wild plants, so you may as well save some money by growing the native varieties. Despite the popular and well-meaning belief that one should not collect seeds from the wild, I can assure you that helping yourself to a few seeds of these species will do no harm. Trust me, I'm a botanist. First, apart from nodding melick they are all common; second, they produce prodigious quantities of seed; and third, they are all long-lived perennials that do not depend on frequent recruitment of new plants from seed. Unfortunately, nodding melick is uncommon in the wild, but is widely available in cultivation.*

*Britain has about 150 native grasses, several of them quite attractive, so there's no need to restrict yourself to the usual suspects listed above. Three more that would make a handsome specimen in a shady spot, yet are hardly in cultivation, are wood false-brome (*Brachypodium sylvaticum*), giant fescue (*Festuca gigantea*) and wood brome (*Bromus ramosus*). All three are common.*

*By the way, although gardeners tend to call almost anything with grass-like leaves and green or brown flowers a grass, many belong in other plant families. Common names don't help – cotton grass (*Eriophorum *species), deer grass (*Trichophorum cespitosum*) and spike rush (*Eleocharis *species) are all in the sedge family. Many true sedges (*Carex *species) make good garden plants; pendulous sedge (*Carex pendula) *is a widely cultivated native, but be warned that it will seed everywhere and its seeds are almost immortal in the soil. Even soft rush* (Juncus effusus), *despite being a bad weed of wet pasture, can look good in the right place.*

Strictly, you should ask the landowner for permission to collect seeds, but since most of the above are plants of moorland or downland, tracking down the landowner may not always be easy. And if you can't identify grasses, buy a picture book, befriend a botanist or better still, don't even try – just grow whatever takes your fancy.

A bit further along the chemical spectrum is the Labiatae (correctly now Lamiaceae), a family with plenty of interesting chemistry but no really poisonous members. This family alone accounts for about half the average spice rack, including mint, rosemary, sage, savory, thyme, hyssop, marjoram, tarragon and basil. Well-known non-culinary members include lavender and catmint, although to my nose the best of all are the small evergreen shrubs in the genus *Prostanthera* (mint bush). As the above list suggests, many members of the family are pleasantly scented, but some have no smell at all and a few are downright horrible. Have a sniff at the leaves of woundwort (*Stachys sylvatica*), a common hedgerow plant throughout the UK, and you'll soon realize why Richard Adams chose General Woundwort as the name of the villain in *Watership Down*.

The chemistry of most Labiatae is relatively uncomplicated, but sometimes investigation reveals hidden depths. If you grow culinary thyme, *Thymus vulgaris*, you will be familiar with its characteristic smell, which comes from essential oils called monoterpenes. What you probably didn't know is that in southern France, where it is a very common wild plant, thyme comes in six varieties, each with its own dominant monoterpene. Two of these, thymol and carvacrol, have the characteristic thyme smell. Oddly enough, although they are chemically different, their smells cannot be distinguished by the human nose. A third, geraniol, smells of lemon, while linalol smells of lavender. The other two are harder to pin down, but alpha-terpineol has a rather sweet 'floral' fragrance. This isn't the limit of thyme's versatility. In Spain there is a seventh variety containing cineol, one of the oils responsible for the scent of rosemary. On the other hand, Spain seems to lack completely the geraniol type.

Scientists have been worrying about why there are so many chemically different thymes for over forty years, and the answer is still not clear. One observation is that the thymol and carvacrol types are the most drought-tolerant, although the role (if any) of the oils in this tolerance is completely unknown. It's worth noting, however, that these and other oils are extremely common in all the small shrubs of hot dry Mediterranean habitats, including lavender and rosemary. Another possibility is that the different oils are there to deter different kinds of herbivores. For example, the sixth French type, containing thujanol-4, grows in the wettest places and is also the type that seems to be the best deterrent to slugs.

Whatever the reasons, thyme's variety has certainly been exploited by gardeners and plant breeders. The linalol type, for example, is sold as lavender thyme. It would also be possible to grow lemon-scented thyme, but in fact the lemon thyme grown by gardeners is a hybrid between *Thymus vulgaris* and *T. pulegioides*, although it's often incorrectly labelled as a separate species, *T. citriodorus*. Other thyme species have smells completely unknown in *Thymus vulgaris*, for example *T. herba-barona*, caraway thyme.

Sadly, the native British flora seems to have missed out on this chemical diversity. As far as I know, our common native thyme, *Thymus polytrichus*, smells only of thyme, and not very strongly at that. Mind you, I still think it's one of our prettiest native plants, and hard to beat for creeping about between paving stones.

At the poisonous extreme is the Solanaceae, which contains perhaps the two most deadly plants in the British flora: henbane (*Hyoscyamus niger*) and deadly nightshade (*Atropa belladonna*). Both contain alkaloids that can cause a variety of symptoms, including coma and death in severe cases. One visible symptom of poisoning by both plants is dilated pupils, and *Atropa* was formerly used by women to make themselves more attractive by dilating their pupils (*belladonna* means 'beautiful lady'). Fortunately neither is likely to be grown in gardens, except by eccentric herbalists, and both are uncommon in the wild, so you're unlikely to encounter either plant. Another member of the same family is tobacco, which of course contains the poisonous alkaloid nicotine, but the garden plant is not likely to cause any harm to humans. (Other well-known alkaloids are quinine (used to treat malaria) and strychnine, but both are extracted from tropical plants (not Solanaceae) that cannot be grown outdoors in Britain.) The most poisonous members of the Solanaceae in temperate gardens are several species of *Datura* (angel's trumpets), either annuals or tender evergreen shrubs with large, dramatic, trumpet-shaped flowers. Given the reputation of the family, it's perhaps not surprising that two edible members, potato and tomato, were both regarded with some suspicion when they originally arrived from America. In fact the tubers are about the only safe part of the potato plant, and even they are poisonous if allowed to turn green, while children have been poisoned by eating potato fruits.

Another family with a surprising mix of edible and poisonous members is the Umbelliferae (now correctly Apiaceae), or carrot family. Fortunately all the members likely to be cultivated are harmless, or indeed edible, e.g. carrot, parsnip, celery, coriander, parsley, fennel, *Astrantia* and *Eryngium*. However, some of the wild British members are very poisonous, for example hemlock (*Conium maculatum*), which resembles a very large cow parsley with a purple-spotted stem. It was used by the ancient Greeks to carry out state executions, Socrates being its most famous victim in 399 BC, and the sixteenth-century English herbalist Henry Lyte described it as 'a naughtie and dangerous herbe'. Until the nineteenth century it was used in medicine, but had an unfortunate tendency to kill the patient. It doesn't sound like a plant you'd want in your garden, but recent research suggests it might be a good slug-killer, or at least a deterrent. I don't have any particular prescription for its use, but you could try strewing the leaves around the plants you want to protect. No need actually to grow it either, since it's a common plant of waste grounds and roadsides.

In practice, even very poisonous garden plants are unlikely to cause any harm, because there's little or no reason to eat most of them. Mind you, I am sometimes surprised by the interesting ways people find to kill themselves. The story of the man who died after stirring his tea with an oleander twig may be apocryphal, but it's certainly true that people have been poisoned by meat roasted on skewers made of oleander wood. In fact oleander is so poisonous I'm surprised there's anyone left alive in the Mediterranean.

The only really dangerous garden plants are those that produce poisonous fruits or large seeds that may attract children. Several species of *Prunus* fall into this category. The fruits, and especially seeds, of many species contain cyanogenic glycosides that release cyanide when acted on by stomach enzymes. The commonest is the cherry laurel, *P. laurocerasus*, but there are reports of children being poisoned by several species. In fact the *seeds* of all woody members of the Rosaceae family, including cherries, apples and almonds, should be treated with care. An American man even managed to die of cyanide poisoning after eating a cupful of apple pips, presumably just to prove it was possible. Other garden plants to keep an eye on if you have small children are cuckoo pint (*Arum maculatum* and *A. italicum*), yew (*Taxus baccata*), snowberry (*Symphoricarpos rivularis*), spindle (*Euonymus* species), *Laburnum*, lily of the valley (*Convallaria majalis*), mistletoe (*Viscum album*), privet

(*Ligustrum* species), pokeweed (*Phytolacca americana*) and *Daphne mezereum*.

A plant worth a special mention, if only because its poisonous principle is one of the most toxic natural substances known, is the castor oil plant, *Ricinus communis*. A tender tree grown in Britain as a half-hardy annual, *Ricinus* is extremely popular as a bold 'tropical' foliage plant, especially in its red, purple and variegated forms. The large, glossy, prettily mottled seeds are very attractive to children, and there are many recent cases of poisoning, nearly all of them children. Just two seeds can be fatal. Ricin, the poisonous compound in *Ricinus*, seems to be a favourite weapon of international terrorists and is also famous for its part in the murder of the Bulgarian dissident Georgi Markov. A novelist and playwright in Bulgaria prior to his defection to Britain in 1969, Markov became a broadcast journalist for Radio Free Europe and had a large audience in Bulgaria. Twice the Bulgarian secret service, assisted by the KGB, tried and failed to assassinate him. But while waiting for a bus near Waterloo bridge in September 1978, he felt a sudden pain in the back of his right leg. As he turned, he saw a man picking up a dropped umbrella. The man apologized in a foreign accent, hailed a taxi and disappeared. By that evening, Markov had developed a fever, and three days later he was dead. An autopsy revealed a tiny metal pellet in the wound made by the umbrella. Examination at the Chemical and Biological Warfare Establishment at Porton Down found that two minute holes had been drilled in the pellet. The holes were empty, but it was later established that they had contained just 450 millionths of a gram of ricin.

GARDENERS' TIP

Don't lose too much sleep over poisonous plants. In the sixteen years between 1962 and 1978 only two people are known to have died from plant poisoning in Britain, one killed by laburnum and the other by the death cap toadstool, which you won't find in your garden. You're far more likely to be struck by lightning or win the lottery, and neither of those has happened yet, has it? Just impress on small children not to put anything they find in the garden in their mouths, and you can grow what you like.

If you're worried about laburnum, grow the sterile hybrid Laburnum × watereri *'Vossii'. It's a better garden plant than the species and produces few if any seeds, so there's also less chance of your garden turning into a laburnum plantation.*

CHEMICAL WARFARE

Poisonous chemicals from plants have long been exploited by gardeners. Rotenone, or derris, is extracted from the tropical climber *Derris trifoliata*. Pyrethrum comes from the flowers of a couple of chrysanthemums that used to be in the genus *Pyrethrum* (hence the name). Both are popular broad-spectrum insecticides which, despite being 'natural' products, are not quite as 'green' as some gardeners believe. However, scientists are only now beginning to investigate some of the more interesting possibilities of such chemicals. Glucosinolates are the chemicals responsible for the taste of brassicas, but they don't just make mustard and horseradish taste interesting. They almost certainly evolved to protect the plants from predators and pathogens, and recent research shows how we can make use of this function in the garden.

Potatoes are host to a range of nasty fungal diseases, including canker and soft rot. Commercially these are controlled by chemical soil fumigants, but new research shows that glucosinolate-containing plants can be even more effective. Almost any brassica will do, but mustard contains some of the highest concentrations. In fact glucosinolates are toxic to a wide range of soil pests, including root-feeding nematodes and the fungus that causes root rot in peas, so mustard may be worth trying against most soil diseases. As a bonus, you can sprout the seeds and eat them too.

GARDENERS' TIP

Mustard is commonly grown as a cover crop or 'green manure', but its value in controlling soil pests depends on getting down into the soil. Sow in early spring, then cut down and add chopped leaves and roots to the top 6–8 inches (20–25 centimetres) of the soil. Mustard grows very fast, so this can be done only a few weeks after sowing.

Before you start, there are three important things to remember about mustard: (a) it's soft and easy to dig in when young, but soon becomes tough if it starts to flower, so don't let it; (b) like all brassicas, mustard is susceptible to club root, so could thwart attempts to control this disease by crop rotation; (c) you need to sow about

3–5 grams of seed per square metre, so buying in 'seed packet' quantities could work out expensive. Look for agricultural seed merchants, where you will be able to buy seed cheaply by the kilo.

BITING THE HAND THAT FEEDS YOU

It's easy to see why plants have poisonous seeds – they don't want animals eating them. It's much less easy to see why plants should have poisonous *fruits*, since fleshy fruits are intended to attract animals to disperse the seeds inside. So why do some plants seem to reward their animal dispersers by trying to kill them?

There is no shortage of plausible explanations for this paradox, but remarkably little conclusive evidence. Some possible reasons are fairly obvious: for example that the toxins are selective and are designed to deter animals that might eat the seeds, but not to harm others that are interested only in the fruit pulp (i.e. good dispersers). This certainly seems to be true in a few cases: the poisonous alkaloid atropine in the ripe fruits of deadly nightshade (*Atropa belladonna*) is fatal to mammals, yet birds seem to eat the fruits without suffering any harm. Actually, even this apparently clear example is not quite as straightforward as it might seem, since it is not clear that mammals would necessarily be poor dispersers of nightshade seeds. Many mammals, such as foxes and bears, enjoy fruit (but not seeds) and are known to be perfectly good seed dispersers.

The logic of other possible explanations is more tortuous.

One possibility is that some chemicals in fruits are intended to speed up the passage of seeds through the gut of potential dispersers, perhaps to reduce damage to the seeds by digestive enzymes. Anyone who was forced to eat prunes as a child will know that fruits can be quite good at this. Anyone unfortunate enough to eat the fruits of buckthorn (*Rhamnus catharticus*) would soon discover that there are much worse things than prunes.

Recently, some particularly inventive ecologists have suggested a completely different explanation. Toxic fruits, at least in the European flora, generally come from toxic plants. In other words, if plants have fruits that make you ill, then the same toxic chemicals are also in the leaves. Maybe plants with toxic leaves are simply unable to keep the toxins out of their fruits, and no special explanation is required for toxic fruits. Of course, this 'non-explanation' begs all kinds of questions. Surely if toxic fruits did prevent or reduce seed dispersal, plants would have been able to do something about it by now? After all, if plants weren't quite good at making their fruits very different from their leaves, then fruits would be tough, green and bitter (which they aren't). Moreover, there is at least one example that shows it's possible to have a toxic plant with non-toxic fruits. All green parts of the yew tree, including the seeds, are highly toxic, yet the bright red aril (not, botanically speaking, a true fruit) is perfectly harmless, if a little insipid.

Sadly, I have to conclude that we actually don't know why some plants have toxic fruits. In fact, there are lots of things scientists don't know, but we manage to keep this embarrassing fact quiet most of the time.

A ROSY FUTURE FOR TOMATOES

One fruit we do know a great deal about is the tomato. Tomatoes are the world's main horticultural crop – every year, three million hectares are grown worldwide, producing 60 million tonnes of fruit. Unripe fruit travel better, so if the crop has to travel before it's sold, there is a strong temptation to pick the fruits while they are still green. At the start of the season, there's another incentive – the first tomatoes of the season usually command a higher price. Final ripening can easily be controlled by manipulating the temperature or by exposure to ethylene gas, which speeds up the process. What's more, tomatoes ripened off the plant are none the worse for it – we've probably all ripened a few green tomatoes at the end of the season. But, there's a problem. The change in colour from green to red is only the final stage of tomato ripening; tomatoes become 'physiologically ripe' while still green. 'Ripe' green tomatoes will turn red, but 'unripe' green tomatoes will not, and there's no way of telling them apart. Even to the most skilled observer, all green tomatoes look the same. For commercial growers, picking green tomatoes that won't turn red is an expensive disaster, since someone has to sort through them all, and worst of all, the unripe tomatoes are wasted.

The change of colour is a combination of the breakdown of chlorophyll (green) and the formation of two carotenoids: *β*-carotene (yellow) and lycopene (red). Scientists have now discovered that the very earliest stages of this process, invisible to the human eye, show up as subtle differences in the infrared light reflected by the fruit, and these changes can be picked up by a sensitive instrument called a spectrophotometer. At the moment the system is too expensive for commercial use, but a suitably developed system could save a lot of wasted tomatoes in the future.

GARDENERS' TIP

If you're left with green tomatoes at the end of the year, you could always just turn them into chutney. But to give them the best chance of turning red, keep them warm (i.e. room temperature, not in the fridge) and put them in a paper bag with a good source of ethylene gas, such as a ripe apple or banana. Light isn't necessary.

A PINKER SHADE OF BLUE

Another aspect of plant chemistry that to some extent runs in families, and that you can actually see, is flower colour. On one level the chemistry of flower colour is simple, since there are just two important families of flower pigments. Carotenoids are much the simpler, and are responsible for yellow- and orange-coloured flowers. Anthocyanins account for everything from scarlet to violet, via pink, mauve, purple and blue. Individual anthocyanins have names that tend to suggest either redness (pelargonidin) or blueness (delphinidin), but of course life is never that simple. The colour of a particular anthocyanin can be changed radically by combination with sugars or other organic molecules, or with metal ions, and by the acidity of the cell sap. In fact exactly the same anthocyanin makes roses red and cornflowers blue. This Jekyll-and-Hyde character of anthocyanins makes it hard to pin down the flower colour blue. Hence the disappointing purplish hue of some Himalayan poppies (*Meconopsis* species), and the fact that some plants, despite passing for blue in the catalogue (e.g. *Clematis* 'Perle d'Azur' and *Hibiscus syriacus* 'Blue Bird'), never quite manage a genuinely aristocratic shade. If you can dimly remember playing with litmus paper at school, you won't need telling how pH affects the colour of anthocyanins. In acid solution they tend to be red, in alkaline solution they tend towards blue. In fact you can make your own acid/alkaline indicator solution from the pigment in red cabbage, which may not entirely justify growing it in the first place, but might just amuse the children one rainy day. Soil pH may also affect flower colour, but not in quite the simple way described above (see Gardeners' Tip on page 55).

The biochemical pathway leading to anthocyanins is rather long and complicated, and many families or genera have a tendency to lack small or large parts of it. Because the gene for a key enzyme, *flavonoid 3, 5-hydroxylase*, is nowhere to be found in the *Rosa* and *Dianthus* genera, it has been impossible to create truly blue roses or carnations by normal plant breeding. However, for those of you who long to own a blue rose, help is at hand. Biotechnologists have isolated the 'blue gene' from petunias and inserted it into both roses and carnations. The not-exactly-blue, but at least violet/mauve carnation 'Moondust' is already in commercial production, and a blue rose cannot be far behind. This does seem to be one genuinely innocuous application of GM technology, since you aren't going to eat blue carnations, and blue roses are unlikely to escape into the countryside. Nevertheless, I find myself strangely unmoved by the whole endeavour. Call me a boring old conservative if you will, but if I want 'petunia blue', I'll grow petunias. Happily, unnaturally blue flowers are only a realistic prospect in plants that already make closely related anthocyanins. Some plants lack so much of the biochemical machinery needed to make anthocyanins that they are unlikely to be persuaded to produce blue flowers. A blue daffodil, thank heavens, is still a long way off.

Trying to persuade plants lacking anthocyanins to make them may be difficult, but persuading plants that already have them to make more is easy. This is just as well, because it has allowed plant breeders to feed gardeners' apparently insatiable appetite for plants with purple, copper or even black leaves. Black plants are, well, the new black, and plant breeders strive to dream up names that emphasize the darkness of their creations: *Physocarpus opulifolius* 'Diabolo', *Heuchera* 'Obsidian', *Phormium* 'All Black', and so on. There's even an International Black Plant Society. I've no opinion either way on such plants, but don't get carried away – overdo it and your garden could literally become a black hole. There's also another problem you may not have thought of: almost 10 per cent of men (but hardly any women) have some form of colour blindness, and the commonest form is deuteranomaly, or 'green-weakness'. If a (male) friend is unmoved by your expensive collection of black plants, that may be because he's deuteranomalous, and instead of all those lovely coppery, purple and wine-red tints, all he can see is something the colour of a cow-pat.

It's worth noting, in passing, that white flowers simply lack both carotenoids and anthocyanins. As all gardeners know, many plants with coloured flowers commonly have white cultivars, so, for example, there are white delphiniums, pelargoniums and foxgloves. Curiously, although nobody knows why, such white varieties are nearly always plants that normally make anthocyanins rather than carotenoids. When did you last see a white dandelion?

GARDENERS' TIP

Those of you of a suspicious nature will already have begun to worry about hydrangeas. As we all know, the flowers of Hydrangea macrophylla *(mophead or lacecap hydrangeas) change colour according to soil pH. Unfortunately, hydrangea flowers are blue on acid soils, pink on alkaline ones – the exact opposite of the reaction to pH described above. This is because the anthocyanin responsible is basically pink, but combines with aluminium to form a blue pigment. Blue hydrangeas therefore need aluminium, but aluminium compounds are soluble only in acid soils. So for a really blue hydrangea, an acid soil is a must. Given an acid soil (pH below 5.5), you can help things along by watering with aluminium sulphate or good old-fashioned alum, but on a neutral or basic soil, blue hydrangeas are impossible. Hydrangea cultivars also differ in their ability to absorb aluminium from the soil, so it pays to grow the right cultivar. Many old hydrangeas (e.g. 'Blue Wave') do not have very brightly coloured flowers and also tend eventually to grow too big for small gardens. A new range of lacecap cultivars from Switzerland, mostly named after birds, are smaller and have much more strongly coloured flowers. On acid soil,* H. macrophylla *'Pfau'* (Peacock) *is intensely blue, while 'Rotschwanz'* (Redstart) *is bright scarlet on an alkaline soil.*

Finally, if your hydrangeas always turn out a depressingly muddy shade of mauve, try growing white cultivars. 'Lanarth White' is very beautiful and the flowers stay pure white, whatever the pH.

BERRIES: SUPER AND OTHERWISE

Not only do anthocyanins look good, they're also powerful antioxidants, and therefore help to combat everything from cancer to bacterial infections. As a result, a whole new industry has grown up in recent years to persuade us to eat (and grow) antioxidant-rich 'superfoods'. As usual, this has all got slightly out of hand. As I write this, flavour of the moment is the blueberry, but such is the speed of foodie fashion, no doubt by the time you read this it will be some hitherto unknown fruit from Tibet or Tierra del Fuego.

In attempting to separate hype from reality, it's worth remarking that we would all be a lot healthier if we grew and ate more fruit and vegetables generally. Nevertheless, fruits do differ a lot in their antioxidant content, so which are the best? Top of the list, out there in a league of its own, is the wild dog rose, but they're fiddly to eat and don't taste great either. You'd be better off drinking rosehip syrup. The dog rose happens to have been tested for antioxidants, but there's no reason to assume it's unique among roses, so you could try eating the hips of almost any rose. The huge hips of *Rosa rugosa* when ripe are really very tasty, although if you eat them all you will have to forgo their ornamental qualities.

Many 'superberries' are plants of northern heathlands, and crowberry (*Empetrum nigrum*), bilberry (*Vaccinium myrtillus*) and cowberry (*V. vitis-idaea*) are all in the top ten. But are they worth eating? Someone thinks they are: a catalogue fell on my doormat recently offering cowberry plants at £8.95 each. Except the marketing department had discovered that the Swedes call them lingonberries, which sounds so much nicer, so that's what they were. They are described as 'tart', which is a bit of a giveaway, since the currants on the same page are described as 'delicious' and 'mouth-watering'. Richard Mabey, in *Food for Free*, says they are 'scarcely edible when raw', but better when cooked. Crowberries, according to Richard, are just plain 'poor eating'. The same catalogue also offers another antioxidant-packed superberry, sea buckthorn (*Hippophae rhamnoides*) at £17.95 a throw. The best the catalogue can say about its berries is that they are 'sharp', so you've been warned. Richard Mabey doesn't mention it at all, and on the evidence of *Food for Free*, he's willing to eat almost anything.

The Swedes, Norwegians and Finns really do eat cowberries and crowberries (and cloudberries, which are basically a miserable, insipid blackberry), so are they mad or what? Well, they genuinely claim to like them, but I think they just don't have much choice. If they could grow dates, mangoes, figs and pineapples, I'm sure they would . . . but they can't.

GARDENERS' TIP

Antioxidants really are very good for you so eat as many as you can. The best 'wild' berry by far is the bilberry, and it tastes great too, so get out in the country and get picking. You can tell how good they are by the colour of your fingers afterwards. Nearly as good are blackcurrants, which taste delicious, crop heavily, and are almost ridiculously easy to grow. I grow 'Ben Sarek', which I strongly recommend. Blueberries are OK, but nowhere near as good as blackcurrants, and not nearly as easy to grow. Two other foods near the top of the antioxidant league are walnuts and pomegranates.

If you still want to try cowberries or crowberries, take a walk in late summer on almost any northern upland, where there are square miles of the wretched things.

ALL CHANGE

Gardening is about change. No two days in the garden are ever the same, and nothing changes more quickly and more dramatically than the colour of flowers, foliage and fruit. Not surprisingly, given the volatility of anthocyanins, red/blue changes are not uncommon among flowers. The lungwort or comfrey family (Boraginaceae) contains an abundance of quick-change artists. Most lungworts (*Pulmonaria* species) have flowers that start out red and become blue as they age. This is the lungwort's way of persuading its pollinators, mainly solitary bees, to visit the younger (red) flowers, which still need pollinating, rather than the blue ones, which have already been pollinated.

The flowers stop nectar production as they turn blue, and bees soon learn that red flowers are worth visiting and blue flowers are not. But why not just let the flowers fade, like most plants? Because the attractiveness of a lungwort plant to bees depends on the total number of open flowers, regardless of colour, so lungworts have cunningly found a way of incorporating older flowers into their overall floral display without allowing these older flowers to deter visiting pollinators.

Autumn leaf colours are one of the principal attractions of some garden plants, and as usual carotenoids and anthocyanins are mainly responsible. All leaves contain carotenoids, although the colour is masked in summer by chlorophyll. As temperatures fall and days shorten, plants gradually stop making chlorophyll and the carotenoids are revealed. Plants whose leaves contain a lot of carotenoids, such as Norway maple, birches and poplars, have the best yellow colours. Anthocyanins, in contrast, are produced in large quantities only in autumn. Because anthocyanins are created by the combination of a pigment molecule and a sugar molecule, their formation is favoured by high sugar concentrations. As the leaf prepares to shut down, an abscission zone forms in the leaf stalk at the point where the leaf will soon fall from the tree. This zone interferes with sugar movement out of the leaf, leading to high concentrations of sugars that react to make anthocyanins, and consequently the characteristic red, pink and purple colours of autumn. Leaves that just turn brown in the autumn, like oaks, contain lots of tannins – the same chemicals that are responsible for the colour and flavour of tea.

The intensity of red autumn leaves, and to a lesser extent yellow, is strongly dependent on the weather. Bright sunlight helps to break down the remaining chlorophyll, and warm, sunny days promote sugar production and anthocyanin formation. Dry weather also increases sugar concentrations, and cool nights help to slow the movement of sugars out of the leaves. Frost, on the other hand, can kill or badly damage the leaves before the pigments develop fully, while dull, rainy weather slows photosynthesis and limits sugar production. Warm, sunny autumn days and dry, cool but not frosty nights therefore produce the most brilliant autumn colours – not to mention the best conditions to enjoy them.

Autumn leaf colours appear to be entirely fortuitous. That is, there's no obvious adaptive value to the plant in having bright autumn colours. It has recently been suggested that they may act as some kind of flag to warn potential leaf-eating insects that the plant is chemically well defended against attack, but the evidence for this idea is so thin as to be almost invisible. Bright colours of *young* leaves, as in *Photinia* × *fraseri* 'Red Robin' and *Pieris* 'Forest Flame', seem a more likely candidate for some overt biological function. We don't know exactly what this function is, but our best guess is that anthocyanins in young leaves act as a kind of botanical sunscreen, protecting leaves from damage by excessively bright sunlight.

Fruits, on the other hand, change colour for fairly obvious reasons. Many turn red to advertise their ripe state to birds, which have particularly acute red vision. As usual, anthocyanins are responsible, and the reaction has much in common with that in autumn leaves, being promoted by high sugar concentrations. In apples the reaction also requires light, which is why only the sunny side of an apple turns red. Of course, not all ripe fruits turn red. The other popular colour for fruits that are dispersed by birds is, surprisingly, black, or shades of dark blue or purple that are close to black. Since fruits are there to be seen, and eaten, by birds, both fruit colours may have evolved to attract birds by standing out against a contrasting background. Red fruits, which are highly conspicuous against a green background, tend to be produced early in the season while leaves are still green. Black, on the other hand, is more obvious against a red, orange or yellow background, so late-fruiting species tend to produce black fruits against a backdrop of autumn leaf colours. Some plants don't rely on leaves to provide the contrast, but let the fruits do the whole job themselves. For example, *Clerodendrum trichotomum* has bright blue fruits set off against a persistent scarlet calyx. In the most shocking clash of colours to be found in the native British flora, spindle (*Euonymus europaeus*) has coral-pink fruits that split to reveal the seeds, each enclosed by a bright orange aril. The cultivar 'Red Cascade' has particularly profuse fruits, but the best display is obtained if it is cross-pollinated by another plant. Both these shrubs are excellent and slightly unusual garden plants.

GARDENERS' TIP

When choosing garden plants, it's easy to get carried away by flowers, and gardening books that arrange plants by flower colour encourage this tendency. Spare a thought, however, for foliage and fruit, which can often provide colour at the beginning and end of the season when flowers are in short supply. Brightly coloured fruits can provide one of the most persistent displays of colour in the garden, and if the birds get them in the end, well, that's all part of the pleasure.

ORIGAMI FOR GARDENERS

In most temperate climates, the biggest and most eagerly awaited season change of all is the annual appearance of leaves in the spring. If, like me, you are lucky enough to have a beech tree or hedge in your garden, you're probably both thrilled and amazed every spring by the speed with which rusty-brown winter colours are replaced by brilliant emerald green ones. It seems to happen almost overnight. Many tree leaves still have quite a bit of growing to do when they emerge from their winter buds, but beech achieves its annual miracle by having the first young leaves almost fully formed, just waiting to unfold. But how, exactly, does beech pack its leaves so neatly into those elegant winter buds?

The key to leaf folding turns out to be the angle between the midrib and the lateral veins. This angle can take almost any value from about 10 to 90 degrees (in other words, the lateral veins can lie almost parallel to the midrib at one extreme, or stick out at right angles at the other). The bigger this angle, the more compactly the leaf can be folded, which means it can be crammed into a smaller bud. However, you rarely get something for nothing in biology, and this economy with space comes at a price. As the leaf unfolds, a smaller vein angle allows the leaf to expand to its full area much more quickly. In beech, which has clearly gone for the rapid expansion option, the angle is towards the lower end of the range, varying from about 50 degrees near the leaf stalk to 30 degrees near the point. The very long winter buds of beech are thus an inevitable consequence of a strategy of fully-formed leaves, folded ready for rapid deployment.

To see how beech leaves are folded, take a sheet of paper about twice as long as wide (10 x 20 centimetres is ideal) and fold it in half lengthways. Mark a point on the folded edge, about 2 centimetres from one end. Draw a line at 60 degrees (or any other angle) from this point and fold along this line. About 1 centimetre further along, mark another line at the same angle and fold in the opposite direction. Keep doing this until you have at least a dozen folds. Now open out along the original long axis fold and, on one half of the paper, convert the valley folds to crests and vice versa. When you've finished, the two halves of your paper 'leaf' should be mirror images. The paper should now fold up just like a real beech leaf. If you try this with a large and a small angle, you should see how the compactness of the folded leaf depends on the vein angle.

You may be alarmed that the academics who discovered this actually got paid for playing with folding paper, but folding biological structures are not just academic curiosities. The whole science of 'biomimetics' uses biological structures to help with human design problems, and the study of leaves can suggest useful ways to design tents and similar structures, and solar panels and antennae for spacecraft. You'll also not be surprised to learn that Japanese scientists are at the forefront of this research, proving that all that time spent inventing origami wasn't for nothing.

Sowing, Planting and Pruning

Danger, Boffin at Work

In my simple-minded view of the world, there's no great mystery about when to sow and plant in the spring. Whether sown outdoors or raised under glass and then transplanted, tender crops (e.g. runner beans and courgettes) are not safe outdoors until the risk of frost has passed. Of course, you do have to decide when this point has arrived.

Hardy crops are a little more problematic (by the way, I should say that although I'm talking about vegetables, exactly the same principles apply to hardy and tender ornamental plants). Seeds of most hardy vegetables germinate very slowly, if at all, below about 5°C, so the usual advice is to aim for a soil temperature of at least 6°C. When does this happen? Well, over most of Britain, the soil is colder than this at the end of February and warmer by the end of the first week in April. Within that period, everything depends on the particular season, the geographical location, altitude and aspect of your garden, how shady or sunny it is, etc. In other words, it's simply not possible to give more specific, general advice. Pessimists could opt to delay sowing until the second week of April every year, and often little would be lost by such a strategy. But in a warm spring you could lose two or three weeks of perfectly good growing weather, and yields of crops that need a long season (e.g. leeks and parsnips) would be lower as a result.

Faced with such uncertainty, some gardeners have sought guidance from unlikely sources, including the moon and stars. In fact, Googling 'planting by the moon' produces over a million hits, so this is clearly not a fringe activity. So, does it work? Before we try to answer that question, a couple of confessions. First I'm basing my interpretation of moon-gardening on the book and website of Nick Kollerstrom, author of one of the bestselling books on the subject. If there are moon-gardeners out there who feel Mr Kollerstrom is misrepresenting their beliefs, then I'm sorry, but please take it up with him, not me. Second, I've no personal experience of planting by the moon, so I've done what I always do when I want to know about a new subject: go to the peer-reviewed scientific literature.

Why the peer-reviewed scientific literature? Let me explain, but before I do, a word of warning is in order: I'm going to give you an insight into the scientific mind at work, and those who regularly consult newspaper astrology columns, or who often use 'nerd', 'geek' or 'anorak' as terms of abuse, should look away now.

Peer review has been the cornerstone of scientific publishing ever since the Royal Society founded the first real scientific journal nearly 350 years ago. It means independent scientific experts vet anything before it appears in print. To pursue a legal analogy, papers in scientific journals are like evidence given in court, given under oath and subject to cross-examination. Everything else, including books, magazines and newspapers, is just hearsay. If you want to write a magazine article or a book claiming that God was an astronaut or that you can lose weight on a diet of cheeseburgers and ice cream, there's nothing to stop you. If you want to say this in a scientific paper, you'd better have some good evidence. This doesn't mean everything in the scientific literature is right, but it does mean that independent experts think your methods were adequate and the data support your conclusions.

Peer review works pretty well, but it isn't perfect. Papers that are wrong, wacky or just plain tedious still get published. However, once they are published, they face an even stiffer test, which is the collective opinion of the scientific community. Good, important papers get mentioned (technically, *cited*) by other authors more than bad, unimportant ones. A really significant paper may be cited hundreds or even thousands of times. Of course, sometimes papers are cited for the wrong reasons. Stanley Pons and Martin Fleischmann's claim to have discovered 'cold fusion', which has now been completely discredited, has been cited hundreds of times, almost always pointing out that the claim is rubbish. However, this is extremely rare – most papers that report things that nobody believes are not refuted, they are just ignored.

So, what does the peer-reviewed literature have to say about planting by the moon? In a word, dear reader, nothing. As far as I can determine, there *is* no scientific literature on planting by the moon. Actually, that's not quite true. There is a review paper by . . . you guessed, Kollerstrom himself. The paper doesn't present any new evidence. It pulls together existing research on the subject and, in one case, also re-analyses some existing data. Kollerstrom's paper *is* in a peer-reviewed scientific journal, but the overwhelming majority of the evidence reviewed in it is *not*. In other words, those who have worked in this area have been unwilling to submit their findings to the editors and referees of scientific journals. Or if they have, their evidence has been found wanting. No doubt you have heard of money laundering. Kollerstrom's article is evidence laundering.

How much notice has been taken of Kollerstrom's paper, i.e. how often has it been cited? The answer is just once, and this citation is in itself extremely instructive. It comes, oddly enough, in a paper in the *American Journal of Obstetrics and Gynecology*, by three doctors investigating the widespread belief that human birth rate is linked to the phases of the moon. Kollerstrom's paper is cited in an introductory paragraph in which the authors list things popularly thought to be influenced by the moon. There it is, rubbing shoulders with the idea that 'Sane people were driven mad by the light of the full moon [and] everyday citizens grew hair and fangs and prowled the night when transformed into werewolves.' So, only one (negative) citation, which is actually quite surprising. Even the most useless papers usually get at least one approving citation, if only in other papers by their own authors. Oh, by the way, is birth rate linked to the phases of the moon? Despite popular belief to the contrary, no, it isn't.

Apart from the lack of published evidence, I have to say there are other things that worry me about moon-gardening. One is the absence of a plausible mechanism. How is the fact that the moon is passing in front of Taurus supposed to improve the germination of my parsnips? Of course, this isn't a fatal problem (where would medicine be if we couldn't use a treatment just because no one knows why it works?), but it does ring alarm bells. We sometimes hear the vague notion that the moon subtly alters the moisture content of the soil, but seriously, British soils in early spring are just plain wet.

Moon-gardening is also quite unlike, say, composting, where a knowledge of the basic principles would enable you to work out the practical details yourself. It has much more in common with a fashionable diet, in that you only know what you're allowed to eat by reading the instruction book. The cornerstone of moon-theory is that there are good and bad days to sow plants grown for their roots, leaves, fruits, seeds or flowers, which sounds simple enough: sow carrots on a root day, lettuce on a leaf day. If only life were so simple. When, for example, do you sow asparagus? Well, Kollerstrom hedges his bets and puts asparagus under roots *and* leaves, which is news to me. If any reader has tried eating either asparagus roots or leaves, I'd be keen to know what they thought of the experience. And leeks? Roots, of course, as are mushrooms, even though fungi are more closely related to you and me than they are to carrots. Cabbage and sprouts are leaves, but broccoli and cauliflower (despite all four being varieties of exactly the same plant) are flowering plants. Kohlrabi, a fifth variety of the same plant, doesn't appear on Kollerstrom's list at all, so if you want to grow that excellent vegetable, you're on your own.

Another striking feature of moon-gardening, which it shares with miracle cures and fortune-telling, is that there's always an excuse when things don't work. In other words, if the magic cure works, then well and good, if it doesn't, it's your fault. For example, apparently it works best 'if crops are not treated with fertilizer'. If, despite abstaining from fertilizers, planting by the moon still doesn't work, then maybe you just did your experiment in the wrong place (one failed experiment took place 'in a quite highly industrialized area of Germany'). Even if you avoid fertilizers and get far away from any industrial influence, there would always be another excuse. In fact Kollerstrom himself makes this quite clear: 'One aims to sow seeds at the peak times of the relevant sidereal energy cycle, yet there is no point in doing so if the ground is too wet, too dry or too cold.'

Finally, I've never seen much point in sowing seeds on a 'leaf day' or a 'root day', since very little happens on the day of sowing. I may sow my parsnips on a root day, but they germinate a week or a fortnight later. Kollerstrom is surprisingly upfront about this too: 'It is crucial to the theory of these experiments that germination occurs on the day of sowing.' Since this has never happened in my garden (or yours), I'm not quite sure where this leaves the whole idea.

GARDENERS' TIP

For the first spring sowings of hardy plants temperature is crucial. You could invest in a soil thermometer, but a simpler solution is to note when the first flush of weed seedlings appears, which indicates the soil is warm enough to sow hardy crops like peas, beetroot and carrots. Later in the season, follow-on sowings can be made at any time. The main hazard for later sowings is lack of water, but there's no need to water seed drills unless it's very dry.

If you have trouble deciding when it's safe outside for tender plants, you could always rely on the old saying:

When mulberry trees are green
No more frosts are seen.

My mulberry tree is in full leaf by mid to late May, so this seems about right. And if you don't have a mulberry tree, maybe it's time you did; it's a beautiful small tree, with quite delicious fruit.

Nor, despite the hocus-pocus about leaf days and root days, is there much difficulty in getting the right balance between roots and leaves. Roots are the plant's way of getting hold of nutrients and water, so if these are in short supply, the plant will go looking for them – by growing more roots. Give a carrot plant too much water and it will reward you with lots of leaves. Too much fertilizer is also to be avoided. On the other hand, any plant that you grow for its leaves or other above-ground parts (e.g. peas, beans and courgettes) will benefit from plenty of water. Potatoes are unusual among root crops in also benefiting from watering, but then potatoes aren't really roots, so there you go.

None of this has anything to do with whether the moon was in Capricorn or not on the day you sowed the seeds.

COMPANION PLANTING

There's no doubt that most plants benefit from the shelter provided by a hedge, that climbing plants like to grow up trees and shrubs, and that shade plants like shade (if only because it stops them being bothered by sun-loving competitors). But none of these is actually a convincing example of a beneficial association between two plants: shelter could be provided by a fence, climbing plants would grow happily up a trellis, and shade could be provided by shade netting or a building. Similarly, other plants will benefit from the atmospheric nitrogen fixed by plants of the pea family, but a sprinkle of nitrogen fertilizer will do the trick just as well.

Generally speaking, plants really hate having to share their space with other plants. As the National Vegetable Society succinctly puts it: 'Plants are not naturally kind to each other.' This dislike of company is so profound that it has left an indelible mark on plant evolution. Seeds have a long list of adaptations designed to prevent germination anywhere near possible competitors. Growing plants can detect potential competitors a long way off, and quickly take appropriate action: they grow rapidly in the opposite direction if it looks like there's any chance of avoiding trouble, and rapidly grow vertically if not.

There are some places where plants like to grow together; in hot dry deserts, it often pays to germinate on the shady side of an established plant (although a convenient rock will do just as well). In the Arctic or on high mountains, growing next to another plant can protect a plant from frost-heaving or bitter, scouring winds. But these are rather unusual places, and if your garden resembles Death Valley or Spitsbergen, no amount of advice from me or anyone else is going to allow you to grow fuchsias or get a decent crop of turnips.

Nevertheless, a widespread belief in companion planting persists among gardeners, and if we pare away the examples (like those mentioned above) that occur only under extreme conditions or don't actually need to involve two plants at all, it's clear that the principal attraction is pest control. So, does it work? Well, we are on much firmer ground than we were with planting by the moon. Companion planting has, especially recently, become a respectable topic of scientific enquiry, and published evidence has begun to accumulate. At what used to be the National Vegetable Research Station, and was then Horticulture Research International, but is now part of the University of Warwick, one team has spent years working on companion planting. It concentrated on pests of brassicas, but it's very likely that the results apply to any winged insect pest looking for somewhere to lay her eggs. As a result, we now understand exactly what happens as a cabbage root fly approaches your cabbage patch, and very illuminating it is too.

1. There's no doubt that pests locate the general area by scent.
2. Once in the right area, pests land on any green object. Smell plays no part at this stage.
3. After landing, pests 'taste' the plant with sensors in their feet. If it's the right plant, they then make a few short flights to other leaves. Every time this also turns out to be the right plant, they become more convinced they've chosen the right spot, and soon they lay some eggs and bang go your cabbages.
4. If they land on the wrong plant, they may leave right away, or try another leaf first. Even if their first landing is on the right plant, if they then land on the wrong plant once or twice, they will most likely give up and move on.

The implications for gardeners are profound. The classic row of vegetables, surrounded by neatly-hoed soil, is an open invitation to pests, since their 'test' flights always bring them back to the same plant. Any background other than bare soil, as long as it's green, will reduce pest attack. In trials, clover or weeds worked just as well as curry plant, marigolds or onions. Indeed, so indifferent are pests to anything other than colour, that model 'companion plants' made of green cardboard were as effective as real ones.

GARDENERS' TIP

Companion planting can be spectacularly effective in preventing attack by insect pests. Anything green will do, including weeds, although this isn't really recommended.

But some words of warning. Get the plant or the timing wrong, and companion planting can do as much harm as the pest. When Gardening Which? *grew cabbage with yellow trefoil, the crop was completely swamped and grew very poorly. Similarly, a companion planting of clover or carrots reduced pest damage to leeks, but again leek yield was low. At least using carrots as a companion plant had the bonus of a crop of carrots . . . The best companions are probably other crops, but these must be carefully matched for vigour, and don't be tempted to plant them too close together.*

Finally, a companion plant with flowers will also attract hoverflies, which may help to control aphids. Nasturtiums would be a good plant to try.

BETTER IN A BED?

Weeds are a major headache for both farmers and gardeners. In fact the fear that weeds might get out of control is a major factor limiting the uptake of organic methods by conventional growers. So if you don't like to use herbicides, what are the options beyond the sensible advice to keep your hoe sharp and use it frequently? This is a problem that preoccupies many agricultural researchers, and some of their findings (though certainly not all) are relevant to gardeners.

A good starting point is that the organic gardener has three main weapons in the war against weeds. We'll come to the third one later, but the first two are the aforementioned hoe and the crop itself: one of the best weed controllers is a healthy and vigorous crop, so more crop equals fewer weeds. But (there's always a but) deploying these two weapons simultaneously is a challenge, because the crop can sometimes get in the way of weeding. If you sow a crop in rows, running a hoe up and down between the rows is easy – it's removing the weeds in the rows by hand that's difficult and time-consuming. But the crop does a better job of controlling weeds if it's sown in uniform blocks, rather than rows.

Some research from Denmark on wheat illustrates this problem. Wheat is normally sown in rows, but sowing in a uniform pattern (or just randomly) covered the ground more completely and therefore was much better at suppressing weeds. Increasing density helped too, and wheat grown at twice the normal density controlled weeds really well, without loss of yield. This is all very well, but relying on the crop to suppress weeds just doesn't work with many crops. Carrots, beetroot or parsnips (however densely and evenly sown) are too feeble to compete effectively with weeds, especially when young. So some weeding will still be required, which brings us back to sowing in rows.

Maybe the answer is to *widen* row spacing and compensate by sowing seeds closer together in the rows (thus keeping crop density per unit area the same), on the principle that the time and effort involved in weeding a metre of row is the same however many plants there are in it. In fact this is just what some recent Dutch research found: increasing row spacing from the standard 0.5m to 0.75m, while sowing seeds closer together to compensate, had no effect on yield and needed less weeding. This was for sugar beet, but the same principles apply to any vegetable sown in rows. If you want to grow only a few lettuces and a patch of radishes, this may all seem academic, but for those with a large vegetable plot or an allotment, the idea of sowing rows further apart, but spacing plants closer within the rows, may be worth trying. If you had to weed a 15-metre row of vegetables, wouldn't you be pleased if you could reduce that to ten metres and still get the same yield?

But there's more than one way to kill a weed, which brings me to weapon number three. Research shows that when it comes to controlling weeds, nothing beats a good old-fashioned mulch. What's more, there's a potential bonus. Almost any mulch will suppress weeds, so newspaper, cardboard, shredded hedge prunings or even plastic are fine. But if you use a mulch that supplies a few nutrients too (e.g. grass mowings or garden compost), this will also boost crop yields without needing to use any added fertilizer. It's also easy and free, although somehow lacking that pleasant 'gardening' feeling that comes from wielding a hoe on a sunny summer's day.

GARDENERS' TIP

Mulches tend to preserve the existing soil conditions, so are best applied when the soil is warm and moist – May is the usual time. Delay if there is a cold snap, and if the soil is dry, wait until after rain. The mulch can then be left undisturbed until the autumn. By then most organic mulches should have mainly disappeared, but the remains can be hoed into the surface soil. Repeat the following May.

NO PRUNING REQUIRED

Pruning, more than anything else, requires the gardener to undertake some serious self-examination. The fundamental question is, what sort of gardener are you? Is your garden mostly a pleasant place to sit, read a good book and watch the world go by? Or do you really enjoy that taming-the-jungle feeling that comes from accumulating a mountain of prunings and then setting fire to it? If the latter, this section is not for you. We are here concerned with those gardeners, the majority I suspect, who prefer a quiet life.

There is probably nothing, other than perhaps the persistent failure of compost heaps to deliver the goods, that strikes more terror into the heart of gardeners than pruning. If you believe the average gardening book, almost every plant requires a different kind of pruning at a different time. In fact, much of this worry can be prevented, because nearly all pruning is unnecessary. There *are* a few plants that really need frequent pruning to give of their best, and we'll deal with those later, but most pruning is a response to a simple problem – a big plant in a small space. This in itself is a consequence of the modern desire for quick results – the leylandii syndrome. With a little foresight and a good dollop of patience, you can have a garden that needs no pruning at all for long periods. The key is simply to grow small plants or slow-growing plants, and to *plant them the right distance apart*. This will have the added benefit of allowing them to develop their proper shape, which in some plants (e.g. *Viburnum plicatum* 'Mariesii' or *Cornus controversa*) is one of the main reasons for growing them in the first place.

SHRUBS THAT NEED LITTLE OR NO PRUNING

Where a genus is listed, the whole genus is intended.

Acer palmatum
Aralia elata
Aucuba japonica
Berberis
Buxus
Camellia
Choisya ternata
Cornus kousa, C. florida
Euonymus fortunei, E. japonica
Genista hispanica, G. tinctoria
Griselinia
Hamamelis
Hebe (especially small species)
Ilex
Kalmia
Leptospermum

Corokia	*Mahonia*
Crinodendron hookeranum	*Magnolia*
Daphne	*Myrtus communis*
Desfontainea spinosa	*Osmanthus* × *burkwoodii*
Elaeagnus	*Paeonia*
Enkianthus	*Pernettya*
Eucryphia	*Pieris*
Pittosporum	*Sarcococca confusa*
Prunus × *cistena, P. tenella, P. lusitanica*	*Skimmia*
Rhododendron (including azaleas)	*Syringa meyeri, S. microphylla, S.* × *persica*
Ribes sanguineum	*Viburnum* × *juddii, V. plicatum, V. tinus, V. davidii, V.* × *burkwoodii*
Rosmarinus officinalis	

Of course, if you plant shrubs that grow slowly, but eventually become quite large, they will look a bit lost when you first plant them. While they are too small to fill their allotted space, fill in the gaps with herbaceous plants or fast-growing shrubs that can be removed later. Also, if you don't find anything you like in the above list, remember that you can almost always reduce or eliminate the need for pruning by careful choice of species and varieties.

Nearly all the common garden genera have members that are considerably smaller than those most commonly found in garden centres. So, if you must have a *Philadelphus* but you only have a small garden, try 'Manteau d'Hermine'. If you like lilac but haven't room for one of the usual

Syringa vulgaris varieties, try *Syringa microphylla* – smaller but just as scented, flowers twice a year and, best of all, doesn't sucker.

FAST-GROWING SHRUBS FOR FILLING GAPS

Again, where only a genus is mentioned, most members are fast-growing.

Abutilon
Amelanchier lamarckii
Brachyglottis 'Sunshine'
Buddleja davidii
Caryopteris × *clandonensis*
Ceanothus
Cistus
Cornus alba, C. stolonifera
Cotinus
Cytisus scoparius
Escallonia
Eucalyptus gunnii
Forsythia
Fuchsia magellanica
Hebe 'Midsummer Beauty'
Hibiscus
Hippophae rhamnoides
Hypericum 'Hidcote'
Kerria japonica
Lavatera olbia
Leycesteria formosa
Lonicera tatarica, L. fragrantissima
Lupinus arboreus
Philadelphus coronarius
Phlomis fruticosa
Potentilla fruticosa
Pyracantha
Rhus typhina
Rubus cockburnianus
Salvia officinalis
Sambucus racemosa, S. nigra 'Laciniata'
Spiraea × *bumalda*
Weigela florida

GARDENERS' TIP

The advice to fill gaps between slow-growing shrubs with faster-growing plants is hardly revolutionary. But, before you set out to do this, make quite sure you know what you're getting yourself into. Eventually your gap-fillers must be sacrificed, so do not become more attached to them than to the prize specimens framing the gap. This is particularly likely to be a problem if more than one person has an interest in the garden. Marriages have foundered on less. A useful precaution is to have an agreed plan of the final shape of the garden that can be used to resolve disputes later. If you don't trust yourself (or your partner), stick to short-lived gap-fillers like Lavandula, Cytisus, Ceanothus, Cistus *or* Lupinus.

SIMPLE AS THAT

Some shrubs really do need routine pruning, however, so let's stop to think why this might be necessary. An almost trivial observation, which nevertheless embraces nearly all you need to know, is that the faster plants grow, the more likely they are to require routine pruning. So, generally speaking, slow-growing shrubs don't need pruning, apart from tidying up or removing dead branches. Another feature of very many shrubs that don't need pruning is flowers produced on old wood in the spring. Think rhododendrons, camellias, magnolias, viburnums and mahonias, for example.

In reality, few garden shrubs need routine pruning, and the ones that do are mostly pretty obvious. *Buddleja davidii* and *Leycesteria formosa*, for example, if pruned hard in spring will produce a relatively large and almost completely new shrub from scratch every year. Leave *Buddleja* unpruned and, although it will continue to flower, it will just get taller and taller until the flowers are out of sight. Because it produces rather soft wood, it will also eventually fall over. Fast-growing shrubs that are routinely pruned to stop them getting tall and leggy usually flower on new wood in the summer.

Two complications must be mentioned. First, don't become so obsessed with pruning that you begin to regard it as an end in itself. Don't lose sight of what pruning is *for*. Nearly always this is to promote flowering, but sometimes the intention is to *prevent* flowering. For example, *Santolina chamaecyparissus* (cotton lavender) is usually grown for its impressive mound of silver foliage, and not for its rather dull flowers. *Santolina* flowers on last year's growth, so the usual practice of pruning hard in spring removes the flower buds and prevents flowering. Sometimes pruning has nothing to do with flowering at all. In *Cornus alba* varieties, bold variegated foliage and colourful young winter stems are promoted by hard spring pruning, and the effect of this treatment on flowering is irrelevant. Much the same applies to *Eucalyptus* (grown as a shrub) and *Sambucus* (elder). Both are often hard pruned in the spring to promote vigorous new growth with impressive foliage.

Second, although the members of the same genus usually behave similarly and require similar pruning, disproportionate confusion is caused by those very few genera where different species require different treatment. For example, the numerous varieties of the common *Hydrangea macrophylla* flower on last year's wood and require little pruning beyond occasional tidying up, but the much more elegant *H. paniculata* flowers on the current year's wood and benefits from hard spring pruning. However, probably more confusion is caused by *Clematis* than by any other genus, although the rules are simple: early-flowering varieties require little or no pruning, while late-flowering varieties are normally pruned hard in the spring. Confusingly, some large-flowered cultivars will flower on old *and* new wood – here the timing and size of the flower displays is up to you. If you really are terrified by the subject of clematis pruning, stick to *C. montana*, *C. alpina*, *C. macropetala*, *C. armandii* and *C. tangutica*, none of which needs pruning.

GARDENERS' TIP

Do not panic about pruning. Get to know the shrubs in your garden and make sure you know when and how they flower. Have a good look at them when they are not *in flower and see what they are up to. Do not be afraid to experiment with pruning to see what gets the best results. The very worst that can happen is that you lose one year's flowers.*

If you find yourself pruning a shrub every year just to keep it within bounds, then it is in the wrong place. Better to move or replace it, but if you must try to keep a large shrub small, prune straight after flowering and remove whole old branches rather than going for a 'short back and sides' approach (however, see Mediterranean shrubs below).

You can also forget everything you have ever been told about pruning hybrid tea roses. There is no need at all to prune every stem back to the traditional 'outward-facing bud', and if you have a lot of roses, they can all be pruned quite easily in five minutes with a hedge trimmer. Nor is hard pruning necessary; never cut back by more than a third, and if you forget completely one year, nothing awful will happen.

On the subject of roses, recent research by the Royal National Rose Society also suggests that dead-heading is best accomplished by doing what the rose would eventually do if left to itself, and snapping the old flowers off at the base of the flower stalk. Roses dead-headed in this way have bigger and earlier second flushes of flower than those dead-headed with secateurs in the traditional way.

Finally, most roses in nurseries and garden centres are grafted, but this is only because it is a cheap way of producing a lot of plants quickly. Roses will grow quite happily on their own roots – just stick some prunings in the ground in autumn and some will have rooted by the following spring.

PRUNING AND TRAINING FRUIT TREES

For ornamental shrubs and trees, a major concern is what they look like after pruning. For fruit trees, however, we have more utilitarian worries – maximizing the yield of apples, plums or whatever. To see why this could be difficult, we have to appreciate that plants' priorities rarely coincide with ours. Plants die if they don't get enough light, so if left alone they tend to grow straight upwards. To make sure of this, the growing tips of branches release hormones that suppress the development of the buds below them. The more vigorous and upright the branch, the more complete is the suppression of other buds on the branch. Botanists call this *apical dominance*, and its practical consequence is no branches, no flowers and no fruit. To persuade some buds to develop into flowers, apical dominance must be broken.

One way to do this, of course, is pruning off the leading shoot. This certainly works; in fact it can work too well. Complete removal of apical dominance often persuades the next bud down the stem to take over and develop into a new leading shoot. Anyone who has pruned an apple tree or wisteria too early in the autumn has seen this happen. A more subtle option is not completely to remove apical dominance, but simply to dilute it. Various forms of training can achieve this, but all involve bending vigorous, vertical shoots into something closer to horizontal. This treatment causes the growing point to produce less hormone, allowing the buds below to develop, hopefully into flowers. Fan-training, espaliers and cordons are traditional approaches, but the limits are set only by your imagination. Apples can be trained into an S shape, while plums can be 'festooned', involving looping young branches over and tying them to the trunk.

There are more drastic options. According to an old proverb, *A woman, a dog and a walnut tree, / the more you beat them the better they be.* You could be forgiven for thinking that our ancestors didn't know much about dogs or women, but they did know a thing or two about walnut trees. 'Beating' involves breaking the branches, but crucially not severing them. The break eventually heals, like a badly set broken bone, but apical dominance is permanently and seriously attenuated. The result is greatly increased flowering and fruiting. Nor does it work only for walnut trees. 'Cracking', or 'brotting', as the Victorians called it, is an effective technique for all fruit trees. It looks a mess, but it works.

GARDENERS' TIP

Shy flowering or fruiting can often be cured by training too-vigorous vertical shoots to the horizontal. It works particularly well for climbing roses, for example. Cracking is also surprisingly effective, but laying into your apple tree with a baseball bat is perhaps best carried out under cover of darkness.

LIFE AFTER PRUNING

Despite the best intentions, many shrubs need pruning eventually. Maybe you have inherited an overgrown garden, maybe you just didn't realize how big some plants were going to grow, or maybe years of neglect have left a shrub looking such a mess that the best course is some drastic pruning. It is at this point that many gardeners begin to worry whether a shrub will survive hard pruning. Fortunately this is one question with a (mostly) simple answer: yes. Few woody garden plants will be killed by hard pruning, and the exceptions are so few that we can afford to deal with them in some detail.

First, a small historical digression. Because botanists know that nearly all woody plants recover from hard pruning, they have never subscribed to the popular myth that Britain's native oak forests were destroyed by shipbuilders or greedy ironmasters. If you cut down an oak wood, all you get, and pretty quickly too, is another oak wood. In fact those who wanted trees for their wood generally helped to protect them from those who wanted not the trees, but the land on which they grew. Once this meant farmers, now it more often means builders of houses, roads, airports or golf courses.

Returning to pruning, nearly all conifers will be killed if cut back to old wood. Indeed, they are remarkably intolerant of pruning. Cut off the top third or quarter of a spruce or fir and the rest will just give up and die. Thus, if you insist on growing an × *Cupressocyparis leylandii* hedge, it must be trimmed to its final size as soon as it gets there. It is no use letting it go for a few years and then cutting the top half off – the result will be a row of miserable fatalities. Just to prove that exceptions have exceptions, none of the above applies to yew (*Taxus baccata*). An overgrown yew hedge can be cut right back to the ground without any fear of permanent damage.

Another group of shrubs that requires careful treatment comes from Mediterranean climates, with cool, damp winters and hot, dry summers (despite the name, such climates are not confined to the Mediterranean basin). A regular feature of such climates is summer fires, and much of the biology of these shrubs has evolved in response to exposure to fires over many millions of years. Essentially the shrubs have adopted one of two possible responses to fire. Some resprout, while others die and regenerate from seed. The resprouters, not surprisingly, don't mind hard pruning, but the seeders have generally abandoned any ability to resprout and die if cut back hard. How do you know which are which? Since some of our favourite garden plants originate from Mediterranean climates, this is an important question.

Fortunately, the distinction is fairly obvious. The sprouters are generally long-lived and are often (but not always) both large and large-leaved. For example, *Laurus* (bay tree), *Arbutus* (strawberry tree), *Rhamnus*, *Osmanthus*, *Phillyrea*, *Daphne*, *Viburnum tinus*. The seeders are smaller and small-leaved. They also tend to be short-lived; in the wild they are accustomed to being incinerated every few years, so they don't see much point in saving up for their old age. Garden examples are *Cistus*, *Lavandula*, *Rosmarinus*, *Cytisus*, *Genista*, *Helianthemum*, *Lotus*, *Phlomis* and *Coronilla*. These are all likely to die if cut back hard. Notice that seeding or sprouting tends to be a generic characteristic – *all* brooms and cistuses are seeders. A few genera break this rule: about half the species of *Ceanothus* (from the Mediterranean climate of California) are sprouters and half seeders. However, there's no easy way to tell them apart, and hybridization has further muddied the waters, so it's best to treat them all as seeders. One final point – fire is not confined to the Mediterranean, but is also a regular feature of north European heathlands. *Calluna vulgaris* and most *Erica* species are seeders.

GARDENERS' TIP

The overwhelming majority of garden shrubs and trees can be pruned very hard and will regrow without any difficulty. Conifers, except yew, will not. Hard pruning of any *short-lived shrub from fire-prone habitats should be regarded as risky. If you must, pruning in the spring and watering well will give you the best chance of success. If in any doubt, take some cuttings as an insurance. Avoid the need for hard pruning by trimming off the current year's growth immediately after flowering.*

Before leaving pruning, a quick word about cuttings. Plants get old and tired, just like people, so to give yourself the best chance of success with cuttings, take them (a) from young, vigorous plants, and (b) when the plant is growing fast, normally in spring or early summer.

THE TRUTH ABOUT HEDGES

As gardening has become more and more a branch of interior decorating, hedges have fallen victim to the modern hunger for quick results. In fact a hedge is much cheaper than a wall, much more robust than a fence, and a better windbreak than either, so anyone establishing a garden from scratch should think seriously about hedging. The ideal hedge plant would grow quickly to about six feet tall and then stop, but unfortunately such a plant does not exist. Shrubs or trees that grow fast when young continue to grow fast when older, so there is an inevitable trade-off between quick results and maintenance requirements. It is easy, if you are in a hurry, to allow the balance to tip too far in favour of rapid growth, but don't forget that a hedge will be around for decades, maybe even centuries. Waiting a few years for your hedge to grow is a small price to pay for less hedge-trimming every year for the next fifty years.

The great majority of countryside hedges consist largely of hawthorn, but this is not a popular hedging plant in gardens, presumably because it looks very bare in the winter and there is no particular need for a garden hedge to be thorny. The ideal hedging plant should be evergreen (or retain its dead leaves in the winter) and grow slowly enough to need trimming only once a year. The longer the hedge, the more important does this latter requirement become, unless of course you actually enjoy hedge-trimming. It is remarkable, given the very few native woody plants in Britain, how many of the very best hedging plants are British natives. Holly, yew and beech are three of the best hedge plants. Holly and yew have beautiful evergreen foliage and are both extremely shade-tolerant. This is important because it means the heavily shaded leaves at the base remain green and healthy. By contrast Japanese privet hedges, which are faster-growing and not so shade-tolerant, have a distinct tendency to become threadbare at the base. A third shade-tolerant evergreen (also a British native) is box. Box grows slowly enough to strain the patience of a saint, but is worth considering on dry chalky soils. Beech is also very shade-tolerant and, while not evergreen, has the curious habit, while young, of retaining its dead leaves all winter. Since annual trimming keeps a hedge in an artificially juvenile state, beech hedges also keep their dead leaves. By the way, *why* juvenile beech (and oak) trees keep their dead leaves is a long-running and unresolved controversy among botanists.

Not a minor consideration when planting a new hedge is cost. Do not even think of buying container-grown plants from the local garden centre. Specialist nurseries (plenty of addresses in any gardening magazine) sell bare-rooted hedging for a fraction of the price. An alternative, if gardening on a truly tiny budget, is to start with holly seedlings. Birds eat holly berries in large numbers and holly seedlings appear in profusion beneath anywhere birds congregate. A good place to look is under a mature tree or large shrub. Since holly seedlings transplant well, this will provide the beginnings of a holly hedge for nothing at all. If you are lucky, and mature yew trees are not too far away, you may even acquire enough yew seedlings for a hedge via this route.

There is a third way to create a hedge, but only if you are blessed with abundant patience and are also prepared to treat gardening as a long-term experimental project. If you erect a post-and-wire fence, given time this will turn into a hedge. Birds will sit on the fence and defecate the seeds they have eaten or drop seeds they are carrying. Such a hedge will certainly be interesting, containing a mixture of everything growing in the surrounding gardens, hedges and woods, but obviously with a bias towards shrubs with berries. Proof that this is an effective way to get a hedge lies in the thousands of miles of hedges that have arisen in North America by this very process. It's slow, but it's absolutely free and entirely effortless.

GARDENERS' TIP

Hedges are not only boundaries, but beautiful garden features in their own right. Among other things, they can provide berries (if not cut too neatly), nesting sites for birds, and shelter for hibernating hedgehogs. Not only that, they are porous at ground level, allowing animals such as frogs, woodmice, foxes and hedgehogs to move freely between gardens. They should therefore be the first choice of boundary in most new gardens. When choosing a hedge, think carefully about future maintenance and not just about how quickly it will reach its desired height. Trimming hedges is much *easier if bordered by a path or short grass, rather than by a flower-bed or by shrubs.*

Those who dislike hedge-cutting can take heart in the recent discovery of an anti-cancer component in yew foliage, which has created a market for yew-hedge prunings. Some drug companies will buy yew prunings, and if your hedge is big enough they will even cut it for you.

Soil and Nutrients

What Are Plants Made of?

PLANTS (OR AT ANY RATE THE WOODY BITS) ARE BASICALLY brown and heavy, and soil is certainly brown and heavy, so it must at one time have seemed entirely natural to think that plants must be made of soil. The eighteenth-century agricultural pioneer Jethro Tull, ahead of his time in many other ways, thought that plant roots had little mouths that ate small particles of soil. This led to ideas about soil cultivation that were absolutely right, but for entirely the wrong reasons. Later, people began to put two and two together and realized that plants couldn't just be made of soil. First they noticed that if you grew a large plant in a tub of soil, once you had got rid of all the roots you were left with the same tub of soil, minus an ounce or two. By 1900 scientists had shown that plants could be grown perfectly satisfactorily with no soil at all, as long as they had water and a few minerals.

We now know that plants are essentially made of air, or at least of carbon, which they obtain from the air (as carbon dioxide) and turn into sugars with the addition of water and the aid of sunlight. However, they also need other things, and by the start of the twentieth century scientists were pretty sure that these other things were nitrogen (N), phosphorus (P), potassium (K), sulphur (S), calcium (Ca) and magnesium (Mg), all in relatively large quantities (hence *macronutrients*), and iron (Fe) in rather small quantities (hence *micronutrients* or trace elements).

The interesting thing about the macronutrients is that two of them (nitrogen and sulphur) have gaseous compounds that enter soil from rain. Nitrogen is also the main constituent of the air and is fixed into a form useful to plants by legumes (plants of the pea family). Potassium, calcium and magnesium are also very common elements in the earth's crust (as is sulphur) and are released by weathering of rocks. The supply of these elements is therefore effectively infinite. Phosphorus is much more of a problem, since it is absent from the atmosphere and relatively scarce in rocks. In fact phosphorus is such a problem for plants that, as we shall see later, getting hold of enough of it has profoundly affected plant evolution since the earliest times.

An interesting question, given that plants survived for millions of years without artificial fertilizers, is how much fertilizer the average garden actually needs. The simple answer is probably none at all. Nutrients are constantly on the move, cycling between soil and plants, but there is no real reason why the total stock of nutrients in your garden should decline. Nitrogen, being a gas, is admittedly a bit hard to keep track of, so the best nutrient to illustrate this principle is phosphorus. Small amounts of phosphorus arrive in your garden in rain, dust and pollen, plus rather more from weathering of minerals in soil. Even smaller amounts are lost in drainage water and by erosion, so generally the natural phosphorus balance is slightly positive.

As long as you don't export large quantities of plant material by, for example, taking wagon-loads of prunings to the tip, things should stay in balance.

The only part of your garden that should lose mineral nutrients on a regular basis is the vegetable plot, from which nutrients are continually removed in the harvested produce. Before you reach for the fertilizer, however, don't forget other nutrient inputs to your garden. The average Briton consumes about 2.2 kilograms of fresh fruit and vegetables every week and, in all but the most self-sufficient of households, almost all is 'imported' from the local supermarket. By my reckoning, about a quarter of this 'consumption' ends up as waste, in the form of banana skins, apple cores, potato peelings, spoiled food, etc. A family of four therefore generates about 114 kilos of plant-based kitchen waste per year, and there's no reason why all this shouldn't find its way on to the compost heap. If we assume that this material contains about 300 milligrams of phosphorus per kilogram (vegetables usually more than this, fruit rather less), then this waste represents about 34 grams of phosphorus. So what does that mean? Well, to put it in perspective, it's equivalent to over 1.5 kilograms of blood, fish and bone, or 773 grams of Phostrogen®, which is 350 gallons at the recommended dilution. Add to that the other 'hidden' nutrient inputs to your garden, including cat and bird food and the nutrients in bought potting compost, plants and seeds, and the lesson is simple: put most of your garden compost on the veg plot and it shouldn't need much else.

GARDENERS' TIP

Most gardens, vegetable plots excepted, need fewer added nutrients than you think. If you recycle your existing stock of nutrients (leave lawn mowings, compost all soft waste and shred prunings or burn them and spread the ash around), there is no reason to add fertilizer routinely to flower beds, shrubberies or lawns except on very poor soils. You can give fertilizer to plants that are regularly hard-pruned (e.g. Buddleja*, roses, many* Clematis*), but remember that they will grow more if you do, so next time you will have even more to prune. Much the same applies to lawns: the more fertilizer you use, the more often you will have to mow. Adding fertilizer to the soil when planting shrubs is a waste of time and may actually harm the plants by discouraging the roots from growing out of the planting hole and into the surrounding soil.*

EVEN MO NUTRIENTS

People have long been aware that plants grown in soil contain more than the six elements listed above, and it gradually began to dawn on them that some of these other elements might be essential for growth. By the 1930s it was obvious that at least copper (Cu), manganese (Mn), zinc (Zn) and boron (B) would have to be added to the list. The conclusive evidence was that a host of frustrating plant diseases, for which no organism seemed to be responsible, could be cured by application of one or other of these four elements. By 1939, people were again happy that the list was complete.

In the 1940s, however, it became apparent that molybdenum (Mo) also was essential, in this case for nitrogen fixation in legumes. Indeed, improvements in yield were so dramatic when tiny amounts of molybdenum were added to deficient Australian soils that one observer commented that you could get more energy out of a gram of Mo than from a gram of uranium. Today we have added cobalt (Co), chlorine (Cl), sodium (Na), silicon (Si) and nickel (Ni) to the essential list, at least for some plants some of the time. The list may not yet be complete, and nor are we entirely sure what some of them actually do.

GARDENERS' TIP

Don't lose any sleep over trace element deficiencies. Such problems are rare on British soils, and if they do occur it is usually soil pH which is at fault (see later).

NUTRIENTS AND PLANT GROWTH

I concluded above that you don't need to put fertilizer on your garden, but if you do, things will generally grow faster. However, this is less than half the story, since different plants vary enormously in their ability to take advantage of extra mineral nutrients. Broadly, evergreen trees and shrubs grow slowly and herbaceous plants grow fast, with deciduous trees and shrubs somewhere in the middle, but there is a lot of variability within each group and overlap between them. Within woody plants, for example, *Buddleja* can grow about ten times faster than slow-growing evergreens such as holly and yew. Elder is another speed merchant, with birch not far behind, while beech isn't much faster than yew. Fast growth, however, comes at a cost. Fast-growing plants like *Buddleja* and elder are not built to last, with soft wood (in fact hardly wood at all) and thin, juicy leaves. This causes all kinds of problems – the wood is physically weak and the leaves are very attractive to herbivorous animals. None of this matters, though, because such plants only do well on fertile soils, so there's no shortage of nutrients to repair any damage. Moreover, these plants are often pioneering colonists of open habitats, doomed to be overtaken eventually by longer-lived and more robust competitors. They therefore don't need to save for their old age, because they don't expect to have one. On the other hand, being built like a battleship, like beech and yew, also carries a penalty. Heavy, dense wood and tough leaves are expensive to produce, so these plants have forfeited the option of growing fast. For yew, slow growth is no longer an option, it's obligatory.

So, what has all this to do with fertilizer? Simply that adding fertilizer to your garden will allow the fast-growers to accelerate, but it won't make much difference to the slow-growers. If mineral nutrients are in short supply, all plants have no option but to grow slowly, but even a hundredfold increase in nutrient supply will only allow yew to grow about 25 per cent faster. Fast growers, on the other hand, may grow five or ten times faster in response to the same fertilizer addition. Lots of fertilizer therefore won't help your slow-growing, 'choice' plants (which also tend to be difficult to propagate and expensive to buy) very much, but it will make fast growers very happy indeed. And not only fast-growing cultivated plants either. Weeds like chickweed, groundsel, bittercress, docks, nettles, thistles and bindweed are all potentially fast-growing and will benefit a lot from extra fertilizer.

A current TV advert for a leading brand of garden fertilizer claims (probably rightly) that it will make your plants grow twice as big. Before you rush out to buy some, consider whether you actually *want* your plants to grow twice as big. An ancient proverb says that it is wise to be careful what you wish for, because you just might get it.

GARDENERS' TIP

Another reason, if you needed one, to go easy on the fertilizer: plants that you thought were growing too fast already will grow faster still. The plants you wish would grow faster will take no notice.

WHERE HAVE ALL THE NUTRIENTS GONE?

'Intensive farming methods have resulted in the decline of the mineral content of our soil – and therefore of what is grown in it.' So complained a recent article in the RHS magazine, *The Garden*. Sounds bad, doesn't it? So what is going on, should we be worried, and what (if anything) can we do about it? Well, in one key respect the *Garden* article has got the cart before the horse. There is no doubt that over the last fifty years, both in the UK and the USA, a wide range of vegetables show significant declines in phosphorus, calcium, potassium, iron, magnesium and copper levels. Given that plants get all these minerals from the soil, it's not surprising that many people have put two and two together and made five. Nor is it surprising that you can now buy bags of rock dust that claim to put all the allegedly 'missing' minerals back into your soil.

To begin with, simple logic suggests that soil is unlikely to be the source of the problem. Vegetables are lower in some minerals that are routinely added to agricultural soils in huge quantities, such as potassium and phosphorous, or which, like iron, are present in all soils in quantities far exceeding the dreams of the greediest plant. Iron is the second most abundant metal on the planet (after aluminium) and makes up fully 5 per cent of the Earth's crust, yet is required by plants in tiny amounts. If you wake up one morning and discover your soil is short of iron, there's only one possible explanation: someone stole your garden during the night. There's also more direct evidence: a national survey of the mineral content of agricultural soils in Britain, begun in 1969, shows not a trace of a decline.

So, the mystery deepens. The minerals are still there in the soil, but not in the vegetables grown in the soil. There seem to be two likely causes of this contradictory state of affairs. The first clue is that the decline has coincided with the widespread adoption and increasing use of massive quantities of chemical fertilizers. Although crops now grow faster and bigger, they do not necessarily manage to take up proportionately more mineral nutrients. This 'dilution effect' means that plants end up with the same concentration of energy, carbohydrate and fat, but proportionately less protein, magnesium and 'micronutrients' such as copper and iron. The same effect even seems to reduce the concentrations of some chemicals that the plants make themselves, such as many vitamins.

But the dilution effect doesn't stop there. Crop breeders may select for many things, but yield is never far from their thoughts. Concentrating on improving yield, uniformity and pest- and disease-resistance can mean that other plant attributes (such as mineral nutrients) gradually decline. In much the same way, roses and sweet peas bred for large flowers and bright colours sometimes lose their scent. A recent study grew several old and new varieties of broccoli together under exactly the same conditions, and found the new, higher-yielding varieties had lower concentrations of several mineral nutrients. Heads of all varieties contained about the same amount of mineral nutrients, but higher-yielding varieties had bigger heads. So the consumer buying a kilo of a bigger, modern variety of broccoli gets the same amount of water, fibre and energy, but less mineral nutrients than they used to.

Crop breeders don't usually try to increase nutritional content, but carrots show what they might achieve if they tried. Compared to fifty years ago, modern carrots contain more than twice the concentration of vitamin A. The reason for this? Growers and consumers both prefer more strongly orange-coloured carrots, and carrot colour comes from ß-carotene, the chemical precursor of vitamin A.

One lesson is clear: crop breeders could easily select for improved nutritional content if the motivation were there, and there are now belated attempts to do this for many of the staple crops of developing countries, including rice, maize, beans and cassava.

GARDENERS' TIP

If you are worried about the nutritional content of the vegetables you eat, what can you do? The first answer is to grow your own and don't overdo the fertilizer. Growing your own has another benefit; mineral contents of vegetables can't change once they've been harvested, but the levels of vitamins and other important chemicals can decrease quite quickly during transport and storage. Your own fresh produce, or that from a local farmers' market, will usually have higher amounts than the supermarket variety.

You should also keep a sense of proportion. Modern vegetables may not be quite as rich in some nutrients as they were fifty years ago, but increasing affluence means we can (and should) easily eat more of them. Also bear in mind that all living things (including plants and people) need more or less the same nutrients in order to grow at all, so no whole-plant food can be seriously short of important nutrients. There is still a vast gulf between the nutritional value of even the limpest supermarket lettuce and highly refined foods like white sugar, rice and bread.

A WORD ABOUT COMFREY

Despite my comments about fertilizers, you might still feel that your garden needs a little extra help, particularly if you have a vegetable plot. How you do this is up to you. You can use chemical fertilizers or you can go for the organic option of blood, fish and bone or seaweed meal. Serious organic gardeners, however, turn up their noses at such unprincipled behaviour – they put their faith in comfrey. Comfrey (specifically, Russian comfrey, *Symphytum* × *uplandicum*) is certainly remarkably rich in the 'big three' mineral nutrients, nitrogen, phosphorus and potassium (NPK), but it is not unique in this respect. Several other plants can do just as well, maybe even a little better, including chickweed (*Stellaria media*), nettle (*Urtica dioica*), sweet cicely (*Myrrhis odorata*), fat hen (*Chenopodium album*), goose grass (*Galium aparine*) and hedge garlic (*Alliaria petiolata*). However, most of these plants are annuals or biennials, which means they never produce enough leaves to be much practical use to the gardener. They also produce a lot of seeds, which means they are likely to make a nuisance of themselves. In fact comfrey has only one serious competitor as a natural fertilizer, and that's nettle. Both are equally fast growing, both are very rich in nutrients, and both will tolerate being cut down several times a year and still come back for more.

Which you grow depends on the value you place on the other things they might do for your garden. Comfrey certainly has a lot going for it. If you cut it often, it will never get round to flowering, but if it does it's a wonderful plant for bees. Comfrey is also one of the most effective herbal remedies, especially for fractures and sprains – my granny used to swear by it. On the other hand, its bristly leaves can cause quite bad dermatitis if you have sensitive skin. Nettle doesn't have any particular medicinal qualities, its flowers are among the dullest in the plant kingdom, and of course it stings, but it does have other qualities. Nettles are the food plant of the caterpillars of several of our more colourful butterflies, including the red admiral, small tortoiseshell and peacock. There is certainly no guarantee that *your* nettles will interest these butterflies, and indeed the latest research suggests they almost certainly won't. To give yourself the best chance, grow your nettles in a sunny spot and cut them down after the first flush of growth – butterflies looking for somewhere to lay their eggs are attracted by fresh young growth. Anyway, even if butterflies don't like your nettles, you can always eat them yourself. So, you pays your money and takes your choice.

Compost activator, liquid feed, mulch – sometimes it seems as if there isn't anything comfrey (or nettle) can't do. In fact organic gardeners are apt to invest comfrey with almost magical properties. However, let's not get carried away. Comfrey is essentially a mechanism for moving nutrients around your garden, but it cannot manufacture them out of thin air, and the NPK in comfrey has to come from *somewhere*. The downside of plants that grow like greased lightning and contain large quantities of minerals is a requirement for a fertile site. You will get the best out of your comfrey only if you can give it a sunny site with deep, moist, fertile soil. Don't imagine that you can grow comfrey in the corner of your garden where nothing else will grow – you know, the bit where the builders left the rubble and subsoil – because you can't.

GARDENERS' TIP

Comfrey really is as useful as they say it is, but it needs looking after. A high-nitrogen fertilizer is ideal – urine will do if nobody is looking. Don't be tempted to cut corners by collecting seed of wild comfrey, which can be invasive and will seed everywhere if allowed to flower. Russian comfrey is sterile, so it won't set seed. 'Bocking 14' is a good cultivar, bred to be especially high in nutrients and tolerant of cutting. Get it from organic gardening specialists. There's nothing wrong with growing wild nettles, and there aren't any cultivars anyway, but if you don't want seedlings everywhere, make sure you grow male plants.

SOIL: MUCK AND MAGIC

The raw material of soil is rock, and how finely this rock is broken up determines what sort of soil you have. Sand grains are about the size of granulated sugar (coarse sand) or caster sugar (fine sand), silt particles about the size of icing sugar, and clay about ten times smaller still. In fact clay particles are so small that their area is enormous – between 50 and 300 square metres per gram of clay. Clay has a bad name among gardeners, but most important soil chemistry takes place on the surface of clay particles, and it is hard to exaggerate its importance to soil fertility. Clay particles have a negative electrical charge, so they tend to hang on to positive ions such as calcium or ammonium, although not very strongly.

Application of lots of ammonium sulphate fertilizer

tends to displace calcium from the clay, and some of this calcium is washed out of the soil and lost. If there is a reserve of calcium in the soil (i.e. lime) the lost calcium is replaced from this reserve and no harm is done, but prolonged application of ammonium sulphate can cause many soils to become acid, or 'sour'. The Park Grass experiment was established in 1856 at Rothamsted in Hertfordshire to examine the effects of different fertilizers on grass yields. Grassland that has received ammonium sulphate without liming for almost 150 years is now down to a pH of 3.7, which is really too acid to grow much except rhododendrons and other acid-soil specialists.

The other major ingredient of soil is organic matter. Dead plant material is broken down by soil animals and micro-organisms, ultimately forming a complex mixture of compounds collectively called humus. These three stores of organic matter (dead plants, soil microbes and animals, and humus) turn over at vastly different rates, the first two much faster than the last. Humus can be carbon-dated like any other organic material and turns out to be remarkably long-lived stuff, persisting in soil for centuries. The slow breakdown of humus thus forms a durable reservoir of mineral nutrients, and in fact it was largely this breakdown that allowed farming to operate at all before the use of artificial fertilizers. Because soils are 'fed' by a constant rain of dead plant material falling on the surface, most of the biological activity in soil takes place in the top few inches. This is true even for very large plants like trees. The next time you see a tree that has fallen in one piece (i.e. not just broken off above the ground), notice how the roots are extensive but surprisingly shallow. Plants may have deep roots that provide anchorage and water, but the really useful roots, in terms of providing the nutrients essential for growth, are close to the surface. This means that when you plant a tree, a wide hole is *much* more use than a deep one.

Careful gardeners devote a lot of effort to maintaining a good soil structure, and rightly so. In a healthy soil mineral particles and humus are stuck together in tiny crumbs a millimetre or two across. These crumbs are held together by electrical attraction between clay particles, but also by the soil inhabitants. A single crumb may contain millions of bacteria, all producing a variety of organic 'glues', plus up to 5 metres of hyphae, the fine threads which make up soil fungi. In a well-structured soil these crumbs are remarkably resistant to damage by, say, heavy rain.

One of the key components of soil is empty space, which may make up 30–50 per cent of a good topsoil. This space is made up of huge numbers of tiny pores and channels, and smaller numbers of larger ones. The smaller pores are usually full of water, the larger ones full of air. Since plant roots need air and water, plenty of pore space is vital for good plant growth. Indeed the right amount and size distribution of pore space is the key to the 'well-drained but moisture-retentive soil' recommended for growing just about anything. A good soil will be free-draining (via large pores and earthworm burrows) and moisture-retentive (owing to a spongy network of tiny pores).

GARDENERS' TIP

It is difficult to overestimate the contribution made to soil structure and drainage by earthworms. A large worm population can be persuaded to do a lot of your cultivation for you: just spread a layer of organic matter on the soil surface in the autumn and let the worms do the rest. Worms like plenty of organic matter and not too much cultivation or pesticides. Oddly enough, worms are particularly badly affected by fungicides, so go easy on them if you value your soil.

If worms like a soil, then so will plants. If you come across plenty of earthworms in your garden, you're obviously gardening properly. If not, you're doing something wrong.

Soil structure is not improved by digging – a last resort for soils that are badly compacted, e.g. by trampling or machinery. If your soil is in good condition, leave well alone.

It sometimes helps to know what sort of soil you have. Discovering a clay soil, for example, may persuade you not to buy a house, or even to move to another part of the country. To find out what sort you have, knead a small amount of moist soil thoroughly in your fingers. If it falls apart, add a little water. If it feels gritty and is difficult to roll into a ball, it's basically sandy. If it can be rolled into a ball, try making a soil 'sausage'. The longer and thinner the sausage you can make without its falling apart, the higher the clay content. Soils with lots of clay can be moulded like plasticine, are shiny when rubbed and sticky when wet.

THE ACID TEST

The amount of a nutrient in a soil is rarely a good guide to how much is actually available to plants. What *is* important is availability, and this is largely determined by pH. pH measures soil acidity on a scale from 1 to 14, but the more extreme values are confined to bottles on the shelf of the chemistry laboratory. Soils range from about pH 3 (very acid) to 8 (alkaline), while 7 is neutral – the pH of pure water. The crucial thing about pH is that at extreme values many plant nutrients are either locked up in insoluble compounds, or alternatively are dissolved in toxic amounts. The pH at which there is neither too much nor too little of anything is around 6.5. Apart from a few extreme acid-soil specialists, *everything* grows best at this pH. Nowhere is this more true than in the vegetable plot, since here you really have very little choice about what you grow. It is remarkable how many common vegetables originate either from lime-rich soils (e.g. parsnips) or from maritime habitats, which also tend to have high-pH soils (e.g. carrots, beet, all brassicas). Not surprisingly, none of these plants will do well in acid soil.

The key problem elements are phosphorus (P), iron (Fe) and manganese (Mn). Phosphorus is in short supply at both extremes of pH, while iron and manganese are unavailable in alkaline soils and present in potentially toxic amounts in acid ones. Since iron and manganese are involved in the manufacture of chlorophyll, a shortage of both is the cause of the lime chlorosis (yellowing of leaves) seen when calcifuge (acid-loving) plants are grown on limy soils. An interesting question, only recently resolved, is how calcicole (lime-loving) plants manage to obtain enough iron and manganese from soils where these elements are effectively unavailable. The slightly surprising answer is that they create their own soil acidity by secreting from their roots large amounts of citric and oxalic acids, the acids of citrus fruit and rhubarb respectively. The former turns out to be the best for dissolving iron and manganese, the latter best for phosphorus.

Calcifuge plants such as rhododendrons normally experience an excess of iron and manganese, so their usual problem is making sure they don't absorb toxic quantities of these elements. They are very good at this, but calcicoles are very poor. Not only do lime-lovers run out of phosphorus in acid soils, they also absorb too much iron and manganese, plus aluminium, which also dissolves in very acid soils. Fortunately, however, these problems only become severe at quite extreme pH, and calcicole garden plants (e.g. rockroses) can safely be grown on moderately acid soils.

GARDENERS' TIP

Get to know the pH of your soil. A chemical soil-testing kit will do this well enough; pH meters sold by garden centres are too unreliable to be much use. If you want the job done more thoroughly, many organizations such as the RHS or Consumers' Association will do a proper analysis. Once you know your pH, it is worth trying to get your vegetable plot as close as possible to pH 6.5. Lime will raise the pH of acid soils, but it is more difficult to lower the pH of alkaline ones. Sustained application of organic matter and ammonium sulphate fertilizer will help, but you will need both perseverance and patience.

In the rest of your garden, do not try to alter the pH. Simply tailor what you grow to your soil. If your soil is limy, learn to love cistuses and viburnums. If you just have to grow rhododendrons, move.

WHO NEEDS ROOTS?

It might seem obvious that plants take up the nutrients they need from the soil through their roots, but unfortunately life isn't always that simple. In fact most plants don't rely entirely, or even mainly, on roots. Instead they have gone into partnership with a type of fungi called *mycorrhizas* (from the Greek for fungus-root). The problem, as usual, is phosphorus. Phosphorus is not very soluble in the soil solution and phosphate ions move extremely slowly through the soil. What this means in practice is that roots can take up phosphorus from the soil much faster than fresh phosphorus can diffuse into the zone around the root. The result is zones around roots that have been sucked dry of phosphorus. One way round this would be for the plants to produce more, finer roots so that every particle of soil has a root next to it, but in reality most plants' answer to this has been to let fungi do the work. Why? Because fungal hyphae are much finer (and hence cheaper) than the finest root; in fact fungal hyphae cost about 1/100 as much to make (in terms of carbon and nutrients), per unit length, as the finest root.

So how does it work? The fungus forms a close association with the root (often penetrating inside) and passes phosphorus to the plant in exchange for sugars. This is a good deal for plants, since they nearly always have plenty of carbon, but hardly ever enough phosphorus. For every metre of root, there may be 100 metres or more of mycorrhizal hyphae, and up to 8,000 metres of hyphae per metre of colonized root has been recorded in some conifers. The problem of getting enough phosphorus must always have been severe for plants, because the fossils of the very earliest land plants, which lived 400 million years ago, show unmistakable signs of the mycorrhizal partnership.

The mycorrhizal association has been so successful that it has arisen several times during evolution, and today there are at least four distinct types of mycorrhizas, involving different groups of plants and fungi. Not the commonest type, but certainly the most conspicuous, is that involving the familiar toadstools. These are found on a range of trees, for instance the well-known fly agaric (*Amanita muscaria*) on birch. Most mycorrhizal fungi do not form such obvious fruiting bodies, but nevertheless the overwhelming majority of plant species (certainly over 90 per cent) are normally mycorrhizal.

It is remarkable that a partnership which is so crucial to plants is almost unknown to gardeners, and indeed to agriculture in general (the culture of truffles, *Tuber* species, is one of the few exceptions). The reasons are simple. Firstly, mycorrhizas are out of sight and therefore out of mind. Secondly, they are so widespread it is difficult to demonstrate their importance by growing plants *without* them. Thirdly, it is much simpler just to apply phosphorus fertilizer than to worry about mycorrhizas. Nevertheless mycorrhizas are important to plants, not just for phosphorus uptake, but almost certainly in other ways that we don't yet fully understand. For example, some research suggests they can protect plants from soil diseases. Unfortunately, mycorrhizas are easily discouraged, particularly by almost every aspect of modern intensive farming. Recent research has found much higher mycorrhizal activity in soil from organic farms compared to soil from conventional farms. Depressingly, mycorrhizas on integrated farms (where chemical fertilizers and pesticides are used only when absolutely necessary, and then in moderation) were just as moribund as those on conventional farms. We now know how easy it is to destroy established mycorrhizal networks, but we've no idea how long it takes to get them back again.

GARDENERS' TIP

All gardeners, and organic ones in particular, can gain much from encouraging mycorrhizas. To see how, follow the directions below.

GOOD FOR MYCORRHIZAS	BAD FOR MYCORRHIZAS
Lots of plants	*Bare soil*
No soil disturbance	*Frequent digging*
Good soil texture	*Compacted soil*
Low rates of fertilizer application	*High rates of fertilizer application*

Slow-release fertilizer
No pesticides

Highly soluble fertilizer
Lots of pesticides
(especially fungicides)

Most garden soils will contain mycorrhizal fungi, or at least their spores, but some new urban gardens may have been created on almost sterile substrates that lack mycorrhizas. Several companies now sell mycorrhizas to add to the soil when planting trees and shrubs on such soils. Try www.planthealthcare.com for one example.

PRETTY PARASITES

One widespread but rather unusual mycorrhizal association is that involving orchids. Orchid seeds are extremely small, so small in fact that they cannot contain any food reserve for the seedling. Orchids are therefore absolutely dependent for their early survival on a mycorrhizal fungus. The odd thing about this association is that nobody has been able to show what the fungus gains from it. Orchids appear to impose on their mycorrhizal 'partners' rather than benefit them. It is a short step from this situation to one in which the plant depends on the fungus for all its food, and many orchids have done just that. For example, the bird's-nest orchid (*Neottia nidus-avis*) has no chlorophyll at all and depends entirely on a *Boletus* fungus. The orchid can now fairly be described as a parasite on the fungus. Other unrelated plants (not orchids) have also adopted the same lifestyle. For example, the yellow bird's-nest (*Monotropa hypopitys*) also depends on *Boletus*. Ironically, *Boletus* has an ordinary mycorrhizal association with trees, so the parasites are ultimately obtaining their carbon from trees. Just to complete an already complex picture, some orchids parasitize the dreaded honey fungus (*Armillaria mellea*).

A surprisingly large number of plants are parasites not on fungi, but on each other. Often this parasitism goes on below ground and is therefore not obvious. This is particularly true when a plant is only partially parasitic (*hemiparasitic* in botanical jargon). Hemiparasites have chlorophyll and photosynthesize normally, but below ground they are attached to the roots of other plants. Some familiar wild flowers, many of them in the foxglove family (although not foxglove itself), are like this. Eyebright (*Euphrasia officinalis*) is one example. The closely related broomrape family contains plants that have gone further and become entirely parasitic. Many broomrapes are choosy about the plants they parasitize and their English and Latin names reflect this. For example ivy broomrape (*Orobanche hederae*) is not uncommon around southern and western coasts, although most broomrapes are rare.

By far the most familiar parasitic plant, however, is mistletoe (*Viscum album*). Legends about mistletoe abound, and one could write a book about this plant alone. Undoubtedly its familiar use at Christmas derives from ancient pagan winter celebrations. It is not hard to imagine the appeal of a plant that sprouted like magic from a tree and, what's more, had evergreen foliage and berries throughout the winter. In fact our mistletoe is just one example of a large family, nearly all tropical, many of which are really quite spectacular plants. Also, while we think of our mistletoe as a rather friendly plant, in many parts of the world, including North America, mistletoes are serious pests of forestry.

GARDENERS' TIP

Because of their unusual nutritional requirements and extremely tiny seeds, hardy orchids have always proved difficult to propagate and are not common in British gardens. This is a pity, since many are every bit as beautiful as their tropical cousins. Expect this to change, however, as new methods of propagation become available.

*The broomrape family contains one of the few wholly parasitic garden plants – purple toothwort (*Lathraea clandestina*). A curiosity rather than really attractive, it is quite easy to grow on the roots of willow or poplar.*

Mistletoe will grow on a range of trees, but is most common on apple. If you want to try growing it, try pushing the seeds into a small nick on the side or underside of a fairly stout branch (at least 3 inches or 7 cm across) at about head height. Before doing so, however, consider the following. Mistletoe berries are not ripe at Christmas, so keep your Christmas mistletoe in water somewhere cool until February or even March. Better still, pick the berries fresh in February. Try not to use berries from bought mistletoe, since much of this is imported from Europe. It may not be suited to our colder climate, or it may even be a different subspecies, common in southern Europe, that grows only on conifers. This will not grow on apple in a month of Sundays.

Finally, remember that mistletoe has male and female plants, so you are not guaranteed to get a female, and what use is mistletoe without berries?

COMPOST

Few things are better for your plants than home-made garden compost, yet few things cause more angst among gardeners than making it. It either refuses to rot down at all, or turns into a slimy mess. Gardening books, which relentlessly emphasize how easy it is to make compost, are not always much help. So, here is what you need to know.

Rapid breakdown of organic matter is caused by bacteria, and it really isn't hard to discover what bacteria like. They like the same things you do. That is, they like to be warm and have plenty of air and enough to eat and drink, but they don't enjoy being wet all the time. Keeping compost warm means it has to be made in a bin, not just a heap. Ideally, an insulated bin. You should also put your compost bin in a sheltered, sunny spot. I know this goes against the grain – *everybody* puts their compost bin in the coldest, dankest and shadiest corner – but you'll just have to decide how much you want to make good compost.

Traditionally, compost is aerated by turning it occasionally, but this is one job nobody enjoys, so it is often neglected. Recent research at the Centre for Alternative Technology in Wales shows, anyway, that turning is unnecessary. The key is to give your compost some structure, so it doesn't collapse into an airless mass. The easiest way to achieve this is by adding paper. Any paper will do, including cardboard, as long as it is first scrunched up and not applied flat. So telephone directories and whole copies of the *Sunday Times* are out (these are ideal for recycling anyway), but breakfast cereal boxes, egg boxes and toilet roll tubes are perfect. Don't add huge quantities of paper all at once, just add it as you go along. But do make sure that large quantities of soft material (e.g. lawn clippings) always have some paper mixed in. The research showed that a compost heap made this way needed no turning at all. It also revealed that a lid was unnecessary, at least in the mild, damp climate of Wales. In drier parts of the country, I suggest you are best watering your heap well (especially after adding dry paper) and then covering it to stop the surface drying out in sunny weather. A lid, which may just be a piece of old carpet, will also help to keep the compost warm.

The need for some structure in your compost heap illustrates perfectly the principle that once a need is perceived, someone will try to make money by filling it. You can now buy 'restructured waste' (aka cornflake packets) at around £3.50 a bag from all good garden centres. However, I'm sure you have enough compostable rubbish of your own, so my advice is not to waste money on other people's.

Although almost anything that used to be alive can be composted, including paper, cotton and wool, some things will break down much more quickly than others. The conventional story, which you will find in most books, is that the key is the carbon/nitrogen (or C/N) ratio. Nitrogen is the 'fuel' of decomposition, while carbon is the padding.

In simple terms, the nitrogen is the roast beef, while carbon is the potatoes and two veg. Material with a low C/N ratio, such as grass cuttings, weeds and vegetable kitchen waste, decomposes quickly, while high C/N materials, such as paper and wood, decompose slowly. There is a lot of truth in this story, and certainly nothing will make wood, on its own, break down quickly. However, there is more to compost than carbon and nitrogen, and recent evidence has highlighted the importance of what chemists call 'bases' – in plant terms, these are calcium, potassium and magnesium. Together these raise the pH, or make plant material less acid, and crucially they also increase the rate of decomposition. The reason is simple: bacteria do not like acid conditions.

Manufacturers of compost activators have always known this, and the major ingredient of the leading brand of compost activator is gardeners' lime, or ground limestone. Lack of bases (especially calcium) is also one of the main reasons that the leaves of some trees make poor compost. Leaves of all conifers, such as spruce, larch and pine, as well as those of oak, beech and sweet and horse chestnut, are low in bases and compost only slowly. Do not make the mistake, however, of assuming all tree leaves make poor compost. Leaves of many trees and shrubs, including elder, laburnum, elm, lime, cherry, sycamore, ash, poplar and willow, are high in bases and make good compost.

GARDENERS' TIP

Keep your compost warm, well aerated and moist but not wet. Mix different ingredients well and add a handful of lime after every 15 centimetres or so. Cover, leave, and enjoy.

Avoid acid tree litter. Oak and beech leaves will *decompose, but the main agents of decomposition are not bacteria but fungi. Fungi work more slowly than bacteria and the product of their labours is leaf mould – good stuff but very low in useful plant nutrients.*

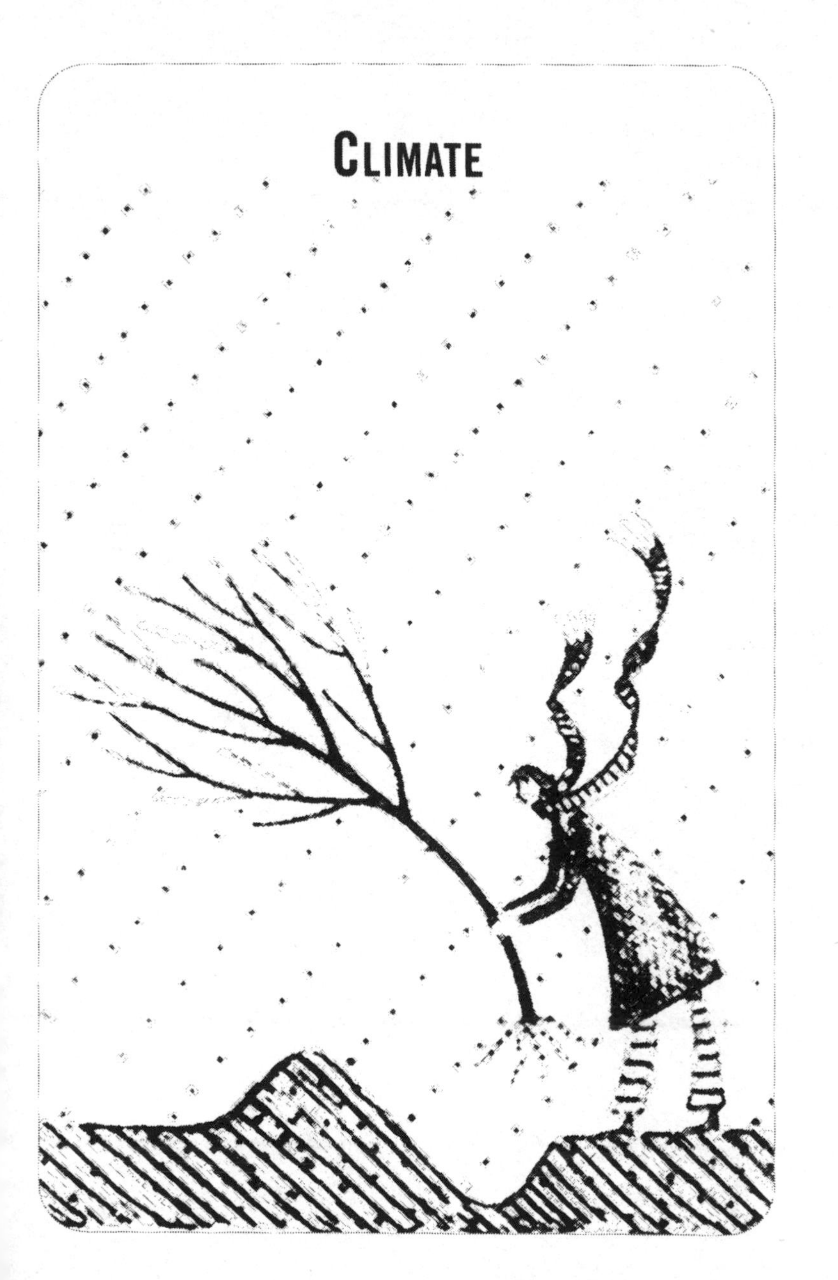
CLIMATE

Is It Hardy?

MANY NATIVE PLANTS HAVE AN IMPORTANT PLACE IN THE garden and some, in particular trees and shrubs, provide a useful framework in many gardens. Nevertheless, the great majority of garden plants are aliens. Some of these aliens are as tough as the natives, but most are not, and we can see the reasons for this if we consider where they come from. We grow very many plants from California, temperate South America, Japan, New Zealand and South Africa. Consider that London lies at a latitude of 51.5 degrees north, and that the corresponding latitudes of Tokyo, Cape Town, Los Angeles, Santiago and Wellington are 36, 34, 34, 33 and 41 degrees north (or south) respectively. The warming influence of the Gulf Stream notwithstanding, it is easy to see, when we also include the many southern European plants in our gardens, that British gardeners are collectively pushing their luck.

Fortunately gardeners can afford to take risks with climate, for at least four reasons:

First, as plants approach their climatic limits, the usual first symptom is sterility. Plants can usually grow beyond where they can reproduce, and so the actual climatic control of distribution operates on seed production rather than on survival. This is of little consequence to gardeners, since they normally don't care whether garden plants produce ripe seed or not.

Second, plants are not killed by average weather, but by extremes. The actual effect of this on natural range limits of plants depends on how long they live. Annual plants may spread well north of their usual limits during a run of mild winters (or hot summers), but winter temperatures that occur only once a century may be enough to limit the distribution of a slow-growing tree. Many gardeners will remember the devastating effect of the unusually severe weather of December 1981 and January 1982. Right across southern and central England, many plants from South America (*Berberis*), California (*Ceanothus*), New Zealand (*Hebe*, *Pittosporum*), Japan or China (*Mahonia*) and southern Europe (*Laurus nobilis*) were cut back to the ground or killed outright. Most gardeners would be happy to put this down to experience and replant. Indeed, many gardeners would grow plants which might be killed once every twenty, or even ten, years. The practical consequence is that many garden plants are routinely grown well outside the climates they could tolerate in the wild.

Third, plants approaching their climatic limits are often killed not by the weather, but by competition from other, better-adapted, species. This fate is particularly likely to overtake alpines that venture into the lowlands. For more on alpines, see below.

Finally, there's climate change. Not so long ago, it would have seemed folly to grow *Callistemon*, *Abutilon vitifolium* or *Phlomis fruticosa* away from the south and west, but many gardeners in the north of Britain now expect plants like these to overwinter, albeit with the benefit of some shelter. In the warmer parts of Britain, gardeners are now chancing a remarkable range of plants, from *Musa* to palms. In terms of the popular USDA plant hardiness zones, the climate change expected in the UK over the next fifty years amounts to about half a zone.

It is sometimes easy to be fooled by an exotic appearance into thinking plants are less hardy than they really are. Cacti are one example. Their exotic blooms and popular association with baking hot deserts all lead one to believe they are tender. In fact cacti are native to just about the whole of North and South America. Most grow in very dry places, but many also experience very cold winters, down to −20°C in the Great Basin desert of Nevada or the Andes for example. No cactus can survive a British winter outdoors, but our wet climate is as much to blame as low temperatures, and many cacti are happy to spend the winter in a cold greenhouse.

GARDENERS' TIP

Don't pay too much attention to catalogue descriptions of hardiness. If you like something (as long as it isn't actually tropical!), have a go. Take sensible precautions, such as avoiding frost hollows, provide good drainage, shelter from east and north winds and mulch roots with straw or compost, and you'll be surprised what you can grow.

ALPINES

High, dry, cold and windy, with pure clear air and endless sunshine. The inability to provide these perceived requirements of alpines deters many gardeners from even attempting to grow them. On closer inspection, however, many plants normally given the rock-garden treatment do not fit this climatic stereotype. Let's look at the evidence:

Many so-called alpines are not plants of alpine habitats at all. They are simply plants of open, rocky habitats at almost any altitude. Familiar British examples, all of which make good garden plants, include thrift (*Armeria maritima*), sea campion (*Silene maritima*), thyme (*Thymus polytrichus*), pasque flower (*Pulsatilla vulgaris*), centaury (*Centaurium erythraea*), maiden pink (*Dianthus deltoides*) and Cheddar pink (*D. gratianopolitanus*). Some of these grow (literally) at sea level on maritime cliffs. Others, like the rockrose (*Helianthemum nummularium*), are so far from being alpine that their real affinities lie in the Mediterranean.

Britain may not be the best place in the world for alpines, but a surprising variety can be found in the mountains of Snowdonia, the Lake District and Scotland. Purple saxifrage (*Saxifraga oppositifolia*), spring gentian (*Gentiana verna*), mountain avens (*Dryas octopetala*), moss campion (*Silene acaulis*) and roseroot (*Sedum rosea*) are all beautiful plants that would grace any garden. Some of these, most famously spring gentian and mountain avens, even grow in luxuriant abundance at sea level in the Burren, a limestone area in the west of Ireland.

Now say what you like about Cornwall, the Lake

District, the highlands of Scotland and the west of Ireland, but they cannot be described as either dry or exceptionally sunny. Nor do they have particularly cold winters. Since that rules out most of the supposed requirements of alpine plants, what are we left with? Not a lot really, except very sharp drainage. The screes, broken rocks and coarse soils of mountains have almost perfect drainage, and as long as you provide this, you can get away with quite a lot. On the other hand, most alpines are not notably tolerant of drought, so moisture retention is also crucial. Providing these conflicting requirements means not even thinking about using ordinary garden soil. Take advantage of raised beds, troughs and pots to create your own soil. Essentially, the wetter your winters, the better drainage you need. A good 'lean mix', suitable for the soggy north and west of Britain, would be equal parts sharp gravel, coarse sand and leaf mould or composted bark (you could use peat, but I very much hope you won't). If you live somewhere drier, you can get away with slightly less sharp drainage – maybe replace the sand with loam.

There are a few other things you can do to keep alpines happy. First, even though you have no control over how much sunshine you get, you can at least make sure nothing gets in its way. So grow your alpines in an open, sunny site – no drips from overhanging branches, no falling leaves, and no shade. Second, because soil pH basically depends on the parent rock, and alpines are in intimate contact with that rock, they are unusually sensitive to soil pH. Most alpines like a neutral to alkaline soil, but some like an acid soil, so make your mind up which you want and act accordingly. Use the right kind of rock (sandstone or granite for acid-lovers, limestone for lime-lovers) and, for lime-hating alpines, water with rainwater, especially in hard water areas. Third, just because alpines look small, don't think you can pack them together like sardines. With alpines, what you see is *much* less than what you get: 90 or 95 per cent of the typical alpine is below ground. So, improve drainage down to at least 30 centimetres and give your plants plenty of room.

Finally, the widely held belief that alpines thrive only beneath a blanket of snow in the winter is a myth. Certainly, snow protects alpines from injury by windblown ice and grit, from desiccation and from injury by very low temperatures, but how many of these things will trouble them in your garden? There *are* alpines that need a covering of snow to keep them dry all winter, but they are in a minority. Unless you want to take alpines really seriously, leave them to the experts.

GARDENERS' TIP

Take account of the above guidance and you'll be able to grow most alpines. Just two final pieces of advice. All that sharp drainage means your plants will need watering in dry spells – never allow the compost to dry out completely. And go very *easy on the fertilizer – alpines are naturally plants of low-nutrient habitats.*

WIND

An important but frequently ignored aspect of climate is wind. Wind certainly affects plant growth, and recent research has shown that this effect operates largely through temperature. Wind evaporates water from leaves and thus cools them – something easy to appreciate if you have ever emerged from a swim in the sea on a windy day. This temperature effect is so strong that in tropical climates windbreaks can kill plants by overheating. In temperate climates, however, windbreaks can increase plant yields by up to 50 per cent. Moreover, there are other benefits. Sheltered plants are less bruised or abraded by wind and are less likely to need staking. Pollinating insects also appreciate shelter.

Agronomists classify crops by how tolerant they are of wind. Some crops, including cereals such as wheat and barley, are highly tolerant. One glance at the East Anglian landscape shows that this lesson has not been lost on farmers, since hedges take up space and therefore cost money. However, *all* the usual garden vegetable crops are in the 'very low tolerance' category, which means that steps taken to increase shelter in gardens will always pay dividends.

GARDENERS' TIP

Books frequently advise that the ideal windbreak should be 50 per cent permeable, but the disadvantages of solid windbreaks are greatly exaggerated. Any *windbreak will provide good protection for a distance of about eight times its height, so a fence or hedge 2 metres tall should be adequate for most gardens. If you can only protect part of your garden from wind, put your vegetable plot in the sheltered part.*

ALL SHOOK UP

Although plants don't like too much wind, they have evolved to cope with its effects by growing shorter, thicker stems. It's not the wind itself that causes this, but the movement of the plant, so exactly the same effect can be produced by shaking or stroking. If you gently stroke or shake a tray of seedlings, they will go on to grow into stockier, more compact and generally stronger plants, which are less likely to keel over in the first squall than plants grown in perfectly sheltered conditions. In fact, of course, it's the plants grown in still air that are abnormally weak and spindly, having been deprived of a stimulus they would normally expect to receive in the real world.

Nor does this apply only to seedlings. Young trees normally flex in the wind, which causes the trunk to grow thicker and stronger. If a young tree is staked too high up the trunk, this flexing can't happen and the tree will be weak. A stake should be no more than one-third the height of the tree. Better to plant a young, unbranched 'whip', which will not require staking. Better still, grow a tree from seed; it's not uncommon for a seedling tree to establish better and eventually overtake a tree planted at a larger size. Absence of a stake allows the movement of the young tree to be transmitted to the roots, which will grow stronger and thicker in the direction of the prevailing wind, bracing the tree against trouble in later life. Given a chance, trees are quite good at avoiding falling over, and most of our attempts to 'protect' them don't really help.

NOT JUST DAFFODILS

Bulbs, and related structures such as corms, are remarkable things; a whole plant in dormant form. Many bulbs need nothing other than water to produce a complete plant. Indeed, daffodils can be forced in almost complete darkness, needing only enough light to persuade them to produce some chlorophyll – a candle is the traditional illumination. No wonder few of us can resist buying a sackful from the garden centre in the autumn, and only later worrying about where to plant them all.

It does pay, however, to give some thought to where bulbs come from. Bulbs are very good at exploiting opportunities for growth that are rather short and rather cool. Short because a plant can be produced from a bulb very quickly, and cool because the inflation of a bulb's contents into a plant (a bit like blowing up a balloon, only with water instead of air) does not need high temperatures, unlike 'real' rapid growth, which does. In the wild, bulbs are therefore common in two quite different sorts of places. First, in deciduous woodland, where they exploit the brief period between the end of the winter and the full opening of the tree leaves. Our native bluebell (*Hyacinthoides non-scripta*) is a good example of this type. Second, in seasonally dry climates, where winters are too cold and summers too dry, but there is a brief window of opportunity in the spring or autumn. This sort of bulb is very common in countries such as Turkey, Greece and Iran. In British gardens we grow both, and knowing the difference will help you get the best out of them.

In bulbs of temperate climates that normally grow in deciduous woodland, as much of the life cycle as possible is condensed into the brief period in spring and early summer before the tree canopy is fully expanded. In daffodils (*Narcissus* species), for example, the formation of next year's flower buds takes place immediately after flowering, while the leaves are dying down. Many other woodland bulbs behave similarly, for example snowdrop (*Galanthus nivalis*) and lily-of-the-valley (*Convallaria majalis*).

Bulbs from drier climates behave quite differently. In many the flower buds form long after the current year's leaves have died. Commercially, this would be after harvesting but before replanting in the autumn. Not surprisingly, flower bud formation requires relatively high temperatures. Tulip (*Tulipa* species), hyacinth (*Hyacinthus orientalis*) and crocus (*Crocus* species) fall into this category. In some others (e.g. bulbous irises) flower buds form during the winter, but only if the bulb has previously experienced high temperatures in the summer.

GARDENERS' TIP

Because daffodils complete all their growth early in the season, they are ideal for growing under deciduous trees or shrubs, and it doesn't matter how deep the shade is later in the year. Tulips are not happy under these conditions, because they don't get the warmth they need to mature next year's flowers. Of course, if you treat tulips as spring bedding, you can grow them anywhere – they will get their warm period after you dig them up. But if you leave them in the ground, tulips will only really thrive in a warm, sunny spot.

Under trees and shrubs, consider growing some of the North American relatives of our own spring bluebells and daffodils. Two in particular, Trillium *and* Erythronium, *are extremely beautiful, and* Erythronium *at least is not difficult to grow.*

DAYLENGTH

How do plants know when to grow, flower or shed their leaves? The timing of growth and flowering is dominated by two things – water and temperature. The trouble with both of these, from a plant's point of view, is that neither is very reliable. As gardeners know only too well, every year is different. Natural selection has therefore favoured plants that set their seasonal clocks by something you can count on, and the solution chosen by many is daylength.

Because daylength varies with season, if you know where you are (relative to the poles), measurement of daylength tells you (and plants) exactly what time of year it is. The importance of daylength declines as you near the equator, where of course it never varies. High latitudes experience the longest days in summer, and also the biggest difference in daylength between summer and winter. Thus many plants from northern climates are 'long-day' plants, i.e. they need long days to flower. Many plants from warmer regions flower in spring or autumn, for the reasons discussed above, and they tend to be 'short-day' plants. Tropical plants tend to be 'day-neutral', a good example of which is the tomato.

Some plants combine daylength cues with temperature cues. Thus the carrot, which like most root vegetables is a biennial, produces a swollen root in its first season and flowers (if we let it) in its second season. Carrots need a period of low temperature, followed by long days, before they will flower, so there is no danger of carrots bolting in their first season. Radishes, on the other hand, need long days to flower, but this does not have to be preceded by a winter, so radishes can (and do) flower in the long days of midsummer.

Commercial growers of pot plants exploit daylength to manipulate their crops. Chrysanthemums are short-day plants, so growers can prevent their stock plants from flowering by keeping them in long days. Cuttings are grown until they reach the desired size and then given short days to initiate flowering. Not surprisingly, those of us without an easy way of manipulating daylength find it hard to make our pot plants do what we want. A particularly frustrating example is the poinsettia (*Euphorbia pulcherrima*), a short-day plant that is brought into flower in time for Christmas. Once it has stopped flowering, days are lengthening and of course it refuses to flower again. Keeping it alive until the following Christmas isn't necessarily the answer either, because normal domestic lighting is bright enough to convince the poinsettia that days are still long. Another plant that often fails to flower when it should if kept too near artificial lighting is the short-day Christmas cactus (*Schlumbergera* hybrids).

GARDENERS' TIP

Don't waste your time trying to preserve the Christmas poinsettia. Apart from the difficulties mentioned above, poinsettias are naturally medium-sized shrubs, and are only persuaded to flower while still so small by spraying with growth regulators which are not available to gardeners.

CLIMATE CHANGE

It's hard to listen to the radio or pick up a newspaper without having climate change drawn to your attention. The weight of opinion now accepts that climate change is a fact, so what can gardeners expect? On one level this a simple question. The latest predictions all agree that by 2080 we can expect average annual temperatures to rise by somewhere between 1 and 5°C, and that this warming will be especially pronounced in the summer in the south and east of Britain. Annual rainfall totals will change little, but there will be big seasonal changes, and again these will be greatest in the south-east. In Cambridge, for example, we expect summer rainfall to decline by almost half, while winter rainfall will rise by a quarter.

So much for the future. If climate change is a reality, we should be able to detect it by now. In fact observations across Europe of the timing of budburst in spring and leaf fall in autumn show that the growing season has been increasing for at least the last thirty years. The growing season has lengthened over this time by eleven days, with over half the change occurring in spring. Observations of 'greening' from satellite and of the timing of seasonal changes in carbon dioxide (which goes down as plant growth starts in the northern hemisphere and then rises again in the autumn) all agree on a seven- to twelve-day increase in the growing season.

What does all this mean for gardeners? Some of what it means can be discerned from changes in behaviour that are already evident. For example, the area of grapevines grown in the UK doubled during the 1990s. In fact growing grapes for wine is still a marginal activity in most years, and the increase may largely reflect increases in tourism and an attempt by existing growers to benefit from economies of scale, rather than any very significant change in climate. Some of it may just be plain old-fashioned optimism. Similarly, the area of forage maize (used for cattle feed), a crop that used to be climatically marginal in the UK, has increased dramatically in recent years. Although the introduction of new varieties better suited to the British climate is responsible for some of this increase, gardeners can expect similar advantages from a warming climate. Crops like sweet corn, which are simply a waste of time at present in much of northern Britain, should be much easier to grow in fifty years' time. Of course, this depends on sufficient summer rain, which means that not just sweet corn, but everything in the vegetable plot may need irrigation in southern gardens in the future.

In fact it may be changes in rainfall that gardeners notice first, especially in the south and east. Lower summer rainfall, combined with higher temperatures, will make it more difficult to grow many large, water-hungry herbaceous perennials. You can expect to find it harder to grow delphiniums, *Anemone* × *hybrida*, *Meconopsis* and peonies, among others. Evergreen shrubs, especially those from drier climates, such as *Ceanothus*, *Cistus*, *Rosmarinus*, *Arbutus*, *Laurus*, *Grevillea*, *Leptospermum*, *Acacia* and *Olearia* may prove much less trouble. Lawns, however, may experience the biggest changes. Shallow-rooted British lawn grasses depend on adequate supplies of water throughout the summer. With water companies increasingly insisting on water metering, gardeners may have some hard choices to make. To give your lawn the best chance of surviving future summers, raise the cutting height on the mower and leave the clippings. Your lawn may still go brown in late summer, but it should recover in the autumn.

Another option, although it's one that may not be necessary until the twenty-second century, is to grow some unfamiliar grasses. Almost all native British grasses are what botanists call C_3 plants, which refers to the sort of photosynthesis they use. The technical details are unimportant, but there is another sort of photosynthesis, called C_4, which works better in warm, dry climates. Many crops of warm climates, including maize and sugar cane, are C_4 grasses. Observations in countries that cross the C_3–C_4 divide, such as New Zealand, suggest that C_4 grasses start to take over once the average temperature of the warmest month exceeds 22°C. The warmest parts of Britain may have crossed this threshold by later this century. There are many C_4 grasses that make adequate lawn grasses, including Bermuda grass (*Cynodon dactylon*), the mainstay of many a Mediterranean lawn. You're never going to make a bowling green out of *Cynodon*, but those whose lawns double as football pitches should get along well enough.

Both computer climate models and recent experience suggest that one unwelcome effect of climate change will be an increase in climatic variability. This means that, paradoxically, our gardens will become both drier (in the summer) and wetter (in the winter), although we can't rule out some intense deluges in summer too (remember summer 2007?). Thus, despite my advice to grow more 'Mediterranean' shrubs, what we really need is plants that are tolerant of drought *and* flooding. Since there aren't many of these, a better option might be to tackle the 'root' cause and take better care of our soils. Many gardeners have become accustomed to relying on chemical fertilizers, with the result that most British garden soils contain too many nutrients but not enough organic matter. Low organic matter makes soils prone to baking hard in summer and compaction and erosion by heavy rain in winter, all problems that are set to get worse. Making and using more compost will improve soil structure, increasing both permeability and water-holding capacity. Other soil-friendly measures we could take include not walking on soil, especially when wet; making sure soil is always covered by growing plants, green manures or an organic mulch; and keeping hard surfaces (paving and decking) to a minimum, since run-off from such surfaces during storms can cause erosion and local flooding.

Finally, climate change will have quite distinct impacts on garden animals. Some of these will be welcome. Small birds are badly affected by hard winters: the wren population, for example, crashed spectacularly after the very cold winter of 1962/3. The recent run of mild winters seems to have allowed wren numbers to increase, although ultimately numbers are limited by other factors such as food supply. Some changes will be less welcome. Aphids, for example, are expected to be more numerous in warmer summers. On the plus side, so are most butterflies. The speckled wood, which arrived in my garden in Sheffield only about five years ago, is continuing to spread north, and is projected to occupy the whole of Britain apart from the Scottish Highlands by the end of the century. Another spectacular success is the comma, whose meteoric expansion seems to be partly due to climate change, and partly due to the switch from a larval food plant that isn't very common (hop) to one that certainly is (nettle). In many cases, we don't yet know how these changes will affect gardeners. Slugs and snails may become more abundant in mild winters, but be less active in dry summers.

GARDENERS' TIP

The future effects of climate change are uncertain, but warmer and drier summers are certain, particularly in the south-east. One sure bet is that you will want to use more water, and that your water company will charge you more for the privilege. If you don't already collect rainwater, buy a butt and drainpipe adapter today.

Also, go easy on the fertilizer. Plants, just like people, soon get used to the good life, and once they have (again just like people) they react badly when the going gets tough. The more fertile your soil, the bigger your plants will grow and the more they will suffer in a drought. In fact, because big, lush plants take more water from the soil, a drought will be worse and start sooner on a fertile soil. Keep your plants a bit hungry and they will look after themselves.

Finally, if you don't already make (and use) garden compost, maybe it's time you did. In the coming struggle with climate change, your compost heap will be your biggest ally.

MAKE COMPOST NOT WAR

In the cause of beating climate change, we are constantly urged to turn down the central heating, use public transport and take our holidays in Yarmouth rather than Yucatan. All good advice no doubt, but is there anything we can do as gardeners to combat climate change? Well, some things are so obvious they hardly need mentioning. Ask yourself if you really need gas-guzzling gadgets like leaf blowers, patio heaters and powered lawn scarifiers. No one with an average-size lawn really *needs* a powered lawnmower. Buy yourself a manual lawnmower instead, which will save you money and help you to stay fit. Some things are slightly less obvious:. nearly 1 kilogram of carbon is burned to make every kilogram of nitrogen fertilizer, and making 1 kilogram of pesticide uses 5 kilograms of carbon. And although it's not strictly gardening, if everyone with a garden hung out a load of washing every week, rather than always resorting to the tumble dryer, we would save half a million tonnes of CO_2 every year.

Carbon neutrality is fashionable, and schemes that allow you to 'offset' your inevitable carbon emissions against carbon-saving activities are sprouting like mushrooms. Much of the money invested goes into planting trees, and you might think that gardeners are in pole position to do their own bit of tree planting. Sadly, however, I have to report that planting trees is a remarkably ineffective way of mopping up surplus carbon. There's about 114 million tonnes (Mt) of carbon in all the UK's vegetation, most of it in trees, which sounds like a lot until you compare it to the UK's annual CO_2 emissions. These vary a bit from year to year, but are generally a little over 150 Mt of carbon per annum. That is, we burn more carbon every year than is contained in all our existing trees and other plants. Put another way, if we could double the area of trees in the UK (in practice, a very tall order), and assuming that trees take around fifty years to grow, we could offset about 1 per cent of the UK's annual CO_2 output. Carbon offsetting by planting trees probably makes a few celebrities and high-profile companies feel better, but as a practical solution to reducing CO_2 emissions, it's not in the same league as buying a few low-energy light bulbs.

Please don't let me discourage you from planting trees, which is a good idea for all kinds of reasons. Urban trees help keep cities cool in summer, reduce dust, noise and pollution, and are great for wildlife. Just don't expect them to make much of a dent in your carbon footprint.

In any case, despite what you read in the newspapers, when it comes to carbon, plants are not where the action is. Globally, there's about four and a half times more carbon in soil than in vegetation, and there would be even more if agriculture hadn't destroyed so much of it. Historically, the world's soils are estimated to have lost around 65 billion tonnes of carbon since the invention of agriculture, and European farmland loses around 100 Mt of carbon every year, in other words an amount not very different from the entire UK CO_2 output. Clearly, the crucial soil carbon battlefield is farming, not gardens, but that doesn't mean we can't do our bit. Garden soils contain too much phosphorus and too much potash (most gardeners use far too much fertilizer), but few have enough organic matter, so make and use as much compost as you possibly can, and don't throw away anything that can possibly be composted. You can't have too much organic matter, so if your vegetable plot has a bare patch, grow a green manure. The great enemy of soil carbon is cultivation, so dig only as a last resort – instead apply compost as a mulch and let the worms do the hard work.

Finally, at the risk of adding to the already deafening chorus of advice on the subject, don't use peat or any peat-based product. We're usually told not to use peat because peat bogs are unique wildlife habitats, but they are also massive stores of carbon. Peat stays as peat because it's waterlogged, but as soon as it's dug up and turned into potting compost, it's well on its way back to the CO_2 and water whence it came.

WEEDS

Is it a Weed?

SOME GARDENS SEEM TO ATTRACT UNWANTED (PLANT) visitors. My mother's is one. As the family botanist, I am sometimes called upon to pronounce on the merits of one of these new arrivals. 'Is it a weed?' To which I can only reply, 'Do you like it?' If the answer to this question is *no*, then it's a weed, if the answer is *yes*, then it isn't. Which brings us to the only worthwhile definition of a weed: *a plant growing where it is not wanted*. Weediness is therefore a value judgement. From time to time ecologists have written volumes, and sometimes held whole conferences, in an attempt to arrive at an objective definition of weediness. All to no avail. The trouble is that you may have a whole garden full of plants that your neighbour would rather die than grow. Because you actually planted all these plants, you may think that means they can't be weeds, but sadly, just because something is planted doesn't stop it being a weed. Some of our worst weeds were originally planted, often in huge numbers.

Many valuable garden plants really are (or were) weeds. Consider corn cockle (*Agrostemma githago*), corn marigold (*Chrysanthemum segetum*) and wild pansy (*Viola tricolor*). At one time or another, farmers have devoted much effort to eradicating all three. In the case of corn cockle, they have very nearly succeeded, and this plant is now close to extinction in the wild in Britain. Yet, to our medieval ancestors, corn cockle was a pernicious weed.

GARDENERS' TIP

*Have the courage of your convictions. Take no notice of what other gardeners think; if you like it, grow it. However, if you take a fancy to hedge bindweed (*Calystegia sepium*), I hope you have understanding neighbours.*

ONE YEAR'S SEEDING, SEVEN(TY) YEARS' WEEDING

Despite the comments above, the classic annual weeds of arable fields, flower borders and vegetable plots have one annoying feature in common: the remarkable persistence of their seeds in the soil. All such weeds have seeds that can live in the soil for at least several decades. It is the absence of this ability that made corn cockle such an easy victim. However, although weed seeds are long-lived, they are not immortal, and they disappear with a characteristic half life of around two years. This means that if you start with 100 seeds of, say, chickweed, after two years you will have only fifty, after four years only twenty-five, and so on. The practical consequence of this is that the number of weed seeds in the soil can be drastically reduced, provided you prevent fresh seed input, but that eradication of weed seeds from the soil can take a very long time.

Note the necessity of preventing new seeds entering the soil. All vegetable crops are most susceptible to competition from weeds when young, and it is easy to show that weeds that germinate once the crop plants are large and well-established have a negligible effect on yield. It is therefore tempting to ignore such weeds, but this is a serious mistake. Even if they do not affect yield this year, they can produce seeds, thus making your job harder *next* year. If you want to reduce your weed problem in the long term, there is no alternative to eternal vigilance. Having said that, you can make your life easier by adjusting crop spacing. Some crops, particularly big leafy ones like cabbages, are pretty good at suppressing weeds, but they can only do this effectively if there aren't big spaces between the crop plants where the weeds can get a head start. So plant crops closer together and in a regular pattern, *not* in rows. Of course, planting in rows makes weeding easier, so you have to be honest and ask yourself how keen you are on weeding. If you're serious about weeding, then plant in rows; if you're not, don't.

Buried weed seeds normally germinate when cultivation exposes them to light. This has led to the novel suggestion that weed problems could be reduced by cultivating at night, but trials have shown that this is not a reliable method of control. Somehow, I don't think it would have caught on anyway.

Many perennial weeds also have very long-lived seeds. Creeping buttercup (*Ranunculus repens*) and docks (*Rumex* species) are good examples. Also, however careful you are, you will always have to put up with wind-dispersed weeds, e.g. willowherbs (*Epilobium* species), groundsel (*Senecio vulgaris*) and dandelions (*Taraxacum officinale*), which will invade from outside your garden.

GARDENERS' TIP

Try to prevent annual weeds from seeding. Remember that many annual weeds can complete their life cycles in only a few weeks, and that weeds hoed out of the ground will continue to ripen seeds in damp weather. Do not waste time hoeing in wet weather, but do it thoroughly when the weather is dry and, even better, windy. You can also prevent weed seeds from germinating by applying a mulch of almost anything: plastic, garden compost, newspaper. Always apply mulches to wet soil. That way they also help to conserve moisture and they don't blow away.

REALLY HORRIBLE WEEDS

Irritating as annual weeds can be, nothing strikes terror into the average gardener's heart like really bad perennial weeds. Horsetail (*Equisetum arvense*), ground elder (*Aegopodium podagraria*), creeping thistle (*Cirsium arvense*), couch (*Elytrigia repens*) and hedge bindweed (*Calystegia sepium*) get around through long underground rhizomes or roots and can regenerate from almost any small piece. Reproduction from seed is not usually important.

The fear these weeds cause is entirely justified; they *are* difficult to control. A key point is that digging them out (unless the infestation is only small) is not usually an option.

Some root too deeply to make this practical and in many cases you could easily end up spreading the problem around. Most often, persistent application of a systemic herbicide, usually glyphosate, is the best bet, but be prepared to go on doing this for years. Spray carefully under shrubs or among bulbs after the foliage has died down. Bindweed can be unwound and spread out on a nearby path or sheet of plastic before spraying. Weeds which are mixed with other plants are almost impossible to control unless you are prepared to paint weedkiller on to individual leaves. Here the simplest option is to remove the plants you want and blanket spray the rest, then replant.

GARDENERS' TIP

Carefully examine new plants to make sure you do not introduce new weeds to your garden. If a perennial weed continually invades from next door, try to interest your neighbour in a joint attack. If you (a) don't like to use weedkillers, and (b) are extraordinarily patient, a certain cure for perennial weeds is to cover the offending area with old carpet for a couple of years.

THE TRUTH ABOUT LAWNS, WEEDS AND MOSS

It is a truth rarely acknowledged, at least by purveyors of grass seed, that lawns are defined by their management. In other words, whatever you start with, from bare ground to a nettle patch, will rapidly become a lawn if it is regularly mown. The reasons for this are simple: only a few plants thrive under frequent close cutting, and most of them are common grasses. If these grasses are around when you start mowing, they will quickly spread, and if they aren't, they will soon turn up. So, if you want a lawn consisting only of fine-leaved grasses (usually fescue, *Festuca rubra*, and bent, *Agrostis capillaris*), then go ahead and buy the appropriate seed or turf. If, on the other hand, all you require is somewhere green and flat to put a deckchair or kick a ball about, then don't bother.

Nevertheless, even if your lawn starts life as seed or turf, it is unlikely to consist only of grass for long. In fact, how realistic is the idea of a lawn containing nothing but various grasses? A lawn is only a domestic version of the sort of short turf produced naturally by sheep or rabbit grazing. Indeed, gardeners transported out of their normal environment into the countryside will often remark on the beauty of natural sheep turf, which only shows that they haven't looked too closely at what they are standing on.

Ecologists, and increasingly the media and general public too, are fascinated by biodiversity. Work on why some sorts of herbaceous vegetation have lots of species and others only a few has revealed an interesting relationship between number of species and biomass, i.e. simply the weight of plant material. This relationship is a curve that is low at each end and high in the middle. That is, vegetation with a high or very low biomass tends to have very few species, while that with an intermediate but lowish biomass often has many species. The key point is that *short turf, including lawns, lies right in the middle of this zone of high potential diversity*. In fact sheep turf on limestone is one of the most diverse sorts of vegetation in the world, sometimes with over forty species per square metre. Not only that, around a quarter of such turf usually consists of mosses and their close relatives, liverworts.

Ecologists still don't agree about why turf of this sort is so exceptionally diverse. Practically, however, there is no doubt both that lawns tend to accumulate species, and that they naturally contain a good deal of moss. Keeping a lawn free of moss and of all but two or three species of grass is therefore the ecological equivalent of trying to push water uphill, i.e. possible but very hard work. Moss can be reduced by good drainage and eliminating shade, and weeds by increasing both acidity and fertility, but probably you will eventually have to resort to weedkillers. Moreover, since a weed-free lawn is in an unstable state, you will have to go on applying weedkillers. The alternative is to accept ecological inevitability and enjoy your weeds. After all, many lawn weeds (e.g. white clover, *Trifolium repens*, lesser trefoil, *Trifolium dubium*, and germander speedwell, *Veronica chamaedrys*) are both attractive and excellent nectar sources for bees.

GARDENERS' TIP

*Try to be relaxed about lawn weeds. Let a patch of lawn grow tall for a while and see what's hiding there. You might be surprised; damp lawns often contain lady's smock (*Cardamine pratensis*), while some southern lawns on chalky soils even harbour the uncommon autumn lady's tresses orchid*

(Spiranthes spiralis*). If you really can't stand unsightly weeds, like plantains or dandelions, tackle them individually with a knife, a spot weeder or by painting with glyphosate.*

Don't panic if your lawn looks like being overrun by moss in winter. Moss is dynamic stuff and always expands a lot during the cooler and damper parts of the year. As long as your lawn is not too damp and shady, the moss will retreat when warmer and drier weather arrives.

And finally, unless you have an unusually large lawn, a manual lawnmower is better for you, the environment and the general peace of the neighbourhood.

FOREIGN INVADERS

'Invasive alien plants are causing havoc in our natural landscape, but gardeners can play a key part in stopping their spread.' At least that's what the exhibit from Imperial College said at the 2002 Chelsea Flower Show. These days it seems that everywhere you turn there's a journalist, a conservation charity or even a scientist telling you to stop growing something in case it escapes and causes mayhem. In one sense this is hardly surprising; if you want to drum up support for a cause or sell newspapers, it always helps to suggest that the country is being overrun by unwelcome foreigners – animal, vegetable or human.

But are things really that bad? Let's begin, as usual, by looking at the facts. It's certainly true that the native British flora is outnumbered by plants from overseas. Only around 1,500 British species are truly native, in that they arrived here without any form of human assistance. In contrast, a huge number of different named plants are available commercially to gardeners, although many of these are cultivars and the number of actual species is much smaller. Ignoring plants that can be grown only under glass, about 14,000 distinct species are currently available. Quite a few more are grown only in botanic gardens, while about another 6,000 alien plants now found outdoors were introduced unintentionally. In all, perhaps 26,000 alien plant species grow somewhere in Britain. Of these something over 1,000 (around 5 per cent) have established self-sustaining populations in the wild. Roughly 70 (0.3 per cent) have made themselves at home to the extent that a botanist might mistake them for natives, about 15 have made a genuine nuisance of themselves, and perhaps half a dozen (0.02 per cent) can be regarded as a really serious problem. In other words, the chance of an alien plant surviving here in the wild at all is slim, and the chance of its becoming a serious pest is almost vanishingly small.

In fact, by international standards, the very low probability of an alien plant becoming a problem in the UK is something of a puzzle in itself. As so often with ecological questions, no one is quite sure why Britain is so resistant to invasion by alien plants, but there are probably at least three contributory factors. First is our mild, damp climate, with no great extremes of temperature or drought. This means every square foot is clothed by a dense, permanent cover of greenery, affording the potential invader no easy toehold. It's no surprise that Mediterranean climates, where seasonal drought regularly opens up the plant canopy, are notoriously vulnerable to invasion. Or that plant invaders often establish initially in highly disturbed urban habitats. Our cool summers also prevent many plants of warmer climate from regularly setting seed in Britain. It's almost certainly this that has prevented false acacia (*Robinia pseudoacacia*) and Japanese honeysuckle (*Lonicera japonica*) from going native here as they have in France and the southern USA respectively. Second is our recent geological history. On several occasions over the last few million years, the British flora has been almost completely wiped out by ice and has had to re-invade from the continent. This has selected a flora capable of rapid spread and establishment in disturbed, relatively fertile habitats – not the sort of plants, in other words, likely to be easily outsmarted by a bunch of newcomers. Third, our plants have a long history of association with people and agriculture, particularly grazing animals. The introduction of European agriculture and animals, which has caused such havoc in other parts of the world, was therefore naturally never a problem in Britain.

Unfortunately, the very same factors that have made Britain so resistant to invasion have made British plants some of the worst weeds in other parts of the world. In New Zealand, for example, the native flora is in serious danger from alien plants, very many of them British. Even apparently harmless British plants, such as mouse-ear hawkweed (*Pilosella officinarum*), have proved to be major problems in New Zealand. It would, however, be a mistake to blame the plants entirely. To a large extent, they are merely filling the vacuum left by the ravages of sheep, goats, deer and rabbits, which have devastated a flora that evolved in the absence of grazing animals. This illustrates the general principle that, although introduced herbivores, predators or parasites frequently cause problems for the organisms they attack, very few extinctions have been caused by the introduction of direct competitors. The many plants introduced to Britain are not known to have caused a single extinction, or even noticeable reduction in abundance, of any native plant. The demise of the red squirrel at the hands of the grey is therefore an ecological curiosity, probably caused as much by a combination of habitat destruction and a parapox virus to which the red is susceptible, as by the grey squirrel itself.

GARDENERS' TIP

There is no need to lie awake at night worrying whether the shrub you just planted is about to set off up the M1, laying waste all in its path. It's extremely unlikely, and in any case, unless you personally import new plants from Nepal or Sichuan, anything you grow is probably already widely grown by others, so the cat is already out of the bag. On the other hand, there's no reason to make life easy for potential invaders. Avoid dumping garden plants in the wild, and if you have surplus invasive garden plants, compost or burn them. This advice applies particularly to water plants, which can spread extremely rapidly in the river and canal network.

PLANTS BEHAVING BADLY

Even though alien plants only rarely cause problems in Britain, the few that do all started out as garden plants. It's not clear whether this is because garden plants are inherently more troublesome than accidental introductions, or simply a reflection of the different modes of arrival. Accidental arrivals have to fight their way ashore against determined opposition from established natives, but garden plants are invited in, given tea and biscuits and asked if they wouldn't like to bring a few friends.

With the benefit of hindsight, the welcome given to some introductions seems surprising. William Robinson, the great Victorian gardener and friend of Gertrude Jekyll, praised its 'graceful arching habit and profusion of small branches of pale flowers in autumn', and recommended 'planting groups of three plants' in gardens or woods. What was he talking about? None other than our old friend Japanese knotweed (*Fallopia japonica*). Japanese knotweed is just one example of garden plants that have escaped from cultivation. As we've seen, most of these fugitives have done little if any harm. Oxford ragwort (*Senecio squalidus*), not to be confused with our native ragwort, *S. jacobaea*, which really is a problem, was cultivated in Oxford Botanical Garden for many years before escaping, and is now a familiar sight on waste land throughout the country. Slender speedwell (*Veronica filiformis*) was imported as a rockery plant, but is now a colourful component of short grass almost everywhere. Whether it is a 'weed' of lawns or not is entirely a matter of opinion – I rather like it. Many other plants (e.g. *Mimulus* species, *Antirrhinum majus*, *Centranthus ruber*, *Oenothera glazioviana*, *Buddleja davidii* and *Cymbalaria muralis*) make a vibrant and inoffensive addition to our flora. These plants are all known to be introduced, but it's remarkable how happy we are to assume a plant is native as long as it behaves itself. The trouble is, the herbalists who often provided the first written records of our plants habitually called a plant 'native' even if they had no real idea whether it was or not and frankly, they didn't much care. This opinion was then passed on from one book to another until it seemed heresy to question it. However, some recent inspired detective work by David Pearman, ex-president of the Botanical Society of the British Isles, has proved beyond reasonable doubt that many attractive 'natives' are in fact ancient garden escapes. Three high-profile examples: snakeshead fritillary (*Fritillaria meleagris*), mezereon (*Daphne mezereum*) and monkshood (*Aconitum napellus*). It's no surprise these are among the most attractive members of our flora: that's why they are here.

Some garden plants, however, have turned out to be real thugs. Apart from Japanese knotweed, two others deserve particular mention. *Rhododendron ponticum* grows in the Caucasus and in Spain, although oddly enough it was native in Ireland in the warm period between two earlier glaciations. Ironically, our plant comes from Spain, where it is now quite seriously endangered. Planted widely as a hardy and reliable flowering shrub, especially in large country gardens, it has now 'gone native' in a big way. It is spreading rapidly, especially in Ireland and the west of

Britain, and has a devastating effect on native plant life – literally nothing can grow underneath it. Seeking an explanation for its very bad behaviour, botanists have recently discovered that our plant isn't even 'pure' *R. ponticum*. It contains a few genes from two other North American species, and although no one knows how this happened, the resulting 'hybrid vigour' may help to explain why it is such a hooligan. Only a little less troublesome is giant hogweed (*Heracleum mantegazzianum*). Kew received seeds of this plant from a Russian botanic garden in 1817, and it was first recorded from the wild in Britain in 1828. This hugely impressive biennial, growing up to twelve feet tall, is now almost the only vegetation along some riverbanks in Britain and Ireland. In addition to causing riverbank erosion and outcompeting native plants, it is also poisonous. The sap can cause severe and painful blistering.

The good news is that the three plants described above are among the very few garden escapes that cause real problems in Britain, although one could add Himalayan balsam (*Impatiens glandulifera*), cherry laurel (*Prunus laurocerasus*) and a handful of water plants to the list.

One glance at giant hogweed and Japanese knotweed and you could be forgiven for thinking that it shouldn't be too difficult to spot potential problem plants a mile away. Both, after all, are about 4 metres tall and look like something out of the pages of *The Day of the Triffids*. You would, however, be quite wrong. Armies of botanists, assisted by national and international organizations, have laboured to find the key to predicting in advance exactly which plants will cause trouble in the future. All this work has been largely in vain. It's not that hard to find plant qualities that are associated with invasiveness in a statistical sense, but these correlations just aren't precise enough to be any practical use. The difficulty, just as with attempts to predict earthquakes, is false positives – you only need to cry wolf wrongly a few times and people stop paying attention. Very much as with people, many factors may predispose someone to criminal conduct, but the only really good predictor of bad behaviour remains a previous offence. A plant that has caused problems in California or Australia is very likely to do the same in New Zealand or South Africa.

GARDENERS' TIP

Don't waste your time growing Rhododendron ponticum. *There are very many other rhododendrons, all of them both more attractive and less trouble. And, although I know* R. ponticum *is very pretty in the spring, the conservation volunteers you may see hacking it down in the countryside deserve your praise and encouragement. They are doing a difficult job that very much needs to be done.*

It is easy to see the appeal of the statuesque Heracleum mantegazzianum, *and I can appreciate why gardeners might want to grow it. If you want something just as impressive but perfectly harmless, try giant fennel* (Ferula communis).

IF YOU CAN'T BEAT 'EM, JOIN 'EM

Most garden plants need some attention, such as staking, dividing or protecting from weeds, if they are to do well. There is, however, a select band of plants that need no such encouragement. It is easy to recognize these plants. They are the ones your neighbours are always offering you a piece of (or, worryingly, a whole bucketful). They are also the plants that eventually take over the gardens of people who don't much like gardening.

Many of these plants are attractive, even beautiful, but growing them entails a continuous battle to keep them under control. The solution is to let them keep each other under control by turning part of your garden into a sort of open prison, where invasive plants can slug it out together without interfering with more delicate specimens. The plan works particularly well for front gardens, which are often small and where you probably don't want to spend much time gardening.

GARDENERS' TIP

The plan has two major variants – herbaceous and woody. It is possible to mix them, but easier to stick to one or the other. The only preparation required is to remove all perennial weeds at the start – the plants themselves will deal with annual weeds and prevent new perennial weeds from invading. Candidate plants are listed in the box, but if you are short of money and have a wide enough circle of gardening friends, you should be able to acquire many of them for free. Once under way, maintenance is zero, although the herbaceous variant can be tidied up with a strimmer at infrequent intervals.

PLANT THUGS FOR THE ZERO-MAINTENANCE GARDEN

HERBACEOUS	
Peruvian lily	*Alstroemeria aurantiaca*
Dwarf bamboo	*Arundinaria vagans*
Michaelmas daisy	*Aster novi-belgii*
Perennial cornflower	*Centaurea montana*
Montbretia	*Crocosmia × crocosmiflora*
Leopard's bane	*Doronicum pardalianches*
Cypress spurge	*Euphorbia cyparissias*
Yellow archangel	*Lamiastrum galeobdolon* 'Variegatum'
Yellow loosestrife	*Lysimachia punctata*
Lemon balm	*Melissa officinalis*
Winter heliotrope	*Petasites fragrans*
Gardener's garters	*Phalaris arundinacea* 'Variegata'
Golden rod	*Solidago canadensis*
Comfrey	*Symphytum* species
Eastern borage	*Trachystemon orientale*

SHRUBS

Ceanothus*	*Ceanothus thyrsiflorus* 'Repens'
Cotoneaster	*Cotoneaster dammeri*
Salal	*Gaultheria shallon*
Persian ivy	*Hedera colchica*
Rose of Sharon	*Hypericum calycinum*
Privet honeysuckle*	*Lonicera pileata*
Oregon grape	*Mahonia aquifolium*
Ornamental bramble	*Rubus tricolor*
Hedgehog rose	*Rosa rugosa*
Greater periwinkle	*Vinca major* 'Variegata'

*Not invasive but spreading

Seeds

Dead, or Just Asleep?

THESE DAYS, OUR INCREASING CONTROL OVER THE WORLD around us leaves us unprepared for seeds that refuse to germinate. Partly this is because domestication of plants that are routinely grown from seed, including all bedding plants, most vegetables and many herbaceous perennials, has removed most traces of seed dormancy. This has not required any conscious selection on the part of plant breeders, gardeners and farmers – however dormant the seeds you started with, many generations of sowing and harvesting has left the awkward customers behind and selected only those individuals that germinate easily. Seed dormancy, however, retains a formidable grip on plants that have undergone little selection by man, and especially among those with long generation times, particularly trees and shrubs. So if you aspire to grow anything a little unusual, anything off the normal well-trodden highway of horticulture, you will sooner or later run up against dormant seeds.

So, first of all, why should seeds not want to germinate as soon as they are shed from the parent plant? This is a question with several answers, but in a cool temperate climate like Britain one answer predominates. Mostly we find winters cold, dark and unpleasant, and so do plants. This applies particularly to seedlings, the most delicate and vulnerable stage in the life cycle, and therefore most plants try to avoid germinating in the autumn or winter. Avoiding germination in winter is easy – all that is needed is a requirement for relatively high temperatures for germination. Avoiding germination in autumn, when conditions are very like those in spring (the preferred season for germination), is much more difficult. Many plants have solved this problem by making seeds that will not germinate until they have been exposed to a period of low temperatures (often this exposure is called *stratification*, but I prefer the term *chilling*). What actually happens during this chilling period is not too important to gardeners, but usually it is one of two things. Seeds of some plants have an immature embryo, which grows to full size during chilling, while seeds of others contain inhibitors that break down or leach out of the seed. Notice that both these processes require the seed to be *imbibed*, or fully hydrated, so keeping dry seeds at low temperatures has no effect on dormancy. Incidentally, botanists have never worried too much about just how cold seeds need to be for chilling to be effective, but anything below about 10°C seems to do the trick.

Gardeners are often recommended to break dormancy by placing seeds in the fridge in a bag of moist perlite, but I tend to disagree, for two reasons. First, chilling seeds require plenty of air, and it is easy to get the medium too wet. Second, the length of chilling period required varies enormously between species, and at the end of it, seeds may germinate immediately at the chilling temperature or they may then need a higher temperature. If they germinate immediately, you must pot them up as soon as possible, and if they don't, you have no way of telling when they have had enough chilling. Far better to arrange for them to let you know, by sowing the seeds in pots of free-draining compost and placing the pots in a cold frame or unheated greenhouse. Then when they germinate you will see them right away, and there is no chance of forgetting the bag of perlite hidden at the back of the fridge behind the remains of the turkey.

If chilling were the only thing seeds might need, life would be comparatively simple. In fact, quite a few plants have gone for a belt-and-braces approach to avoiding autumn germination, requiring a warm period followed by chilling. Again, several different things can happen during these two periods. Sometimes an immature embryo grows during the warm period and inhibitors leach out during the cold period. *Lonicera fragrantissima* is a good example. In roses and many other woody members of the rose family, including some cherries and hawthorns, inhibitors can leach out only after the tough woody seed coat has split during the warm period. This splitting is largely caused by bacterial decay and can be greatly accelerated by adding a compost activator to the sowing medium.

Some plants have worked a neat variation on the chilling theme. Although it seems obvious that seeds don't want to germinate immediately before the winter, what they really want to avoid is being *above ground* during the winter. Roots are not at great risk at this time, and it could even be advantageous to get a head start by having a root system ready to go in early spring. So in some plants the 'top' half of the seed is dormant and the 'bottom' half is not. Botanists call this 'epicotyl dormancy'. These seeds germinate as soon as they are sown, but produce only roots at first; the shoot grows only after a period of chilling. Examples among garden plants include bugbanes (formerly *Cimifuga*, now *Actaea*), *Hepatica* and all species of *Viburnum*.

In either of the above two cases, the best way to deal with dormancy is to sow seeds as soon as possible. This means as soon as they are ripe if collecting your own seed, or as soon as they arrive if you buy them. Either way there is a potential problem, because the warm and cold periods must occur in the right order. So, if you miss the first warm period, nothing happens during the first winter, and the seed takes two years from sowing to germinate. A few uncooperative plants, including wake robin (*Trillium grandiflorum*) and bloodroot (*Sanguinaria canadensis*), have such complex dormancy that they never germinate until after experiencing two winters.

Although a need for winter cold is the main cause of dormancy in British garden plants, we also grow a significant minority with so-called 'hard' seeds. That is, seeds with coats that are completely impervious to water, so that the embryo inside is dry. These seeds rarely have any other sort of dormancy, so that germination takes place as soon as the seed coat is breached and water enters. Here dormancy is best broken by cutting the seed coat with a razor blade or rubbing the seed gently between two sheets of fine sandpaper. Seeds in which this treatment has been effective should visibly swell when water is added. Hard seeds are very much a family characteristic and are common in the

Leguminosae, Malvaceae, Convolvulaceae, Cistaceae and Geraniaceae.

A few final pieces of advice. The flesh of berries often contains germination inhibitors, so always remove this completely before sowing. If this is difficult at first, soak the fruits in water for a few hours or days until it becomes really soft. Finally, remember that we have dealt so far with outdoor garden plants. If you have any difficulty germinating tropical house or conservatory plants, chilling is not the answer.

GARDENERS' TIP

In growing your own plants from seed, as in so much else in gardening, patience is a great virtue. If you are dealing with seeds that you suspect are dormant, do not give up hope until at least the second spring after sowing. Unless the seeds were dead to begin with, mother nature should eventually provide them with what they need, even if you are not too sure what that is.

AFTER DORMANCY

Even after dormancy has been overcome, or if your seeds were not dormant in the first place, they still need to be persuaded to germinate. This is rarely as tricky as breaking dormancy, but it still pays to know a few simple principles. To a large extent, plants germinate in response to light and temperature (but see below), and of these light is much the simpler. Most seeds will germinate in darkness, and some even benefit from it, but small seeds (say, lobelia or foxglove size or smaller) need light and should not be covered with compost when sown.

Temperature is much more interesting. First, something so obvious you probably think it doesn't need saying – tropical and subtropical plants germinate best at high temperatures. If you're concerned only with house or conservatory plants, this is easy, since you germinate and grow these plants indoors. Don't forget, however, that we also grow vegetables from warm climates outdoors, e.g. sweet corn (*Zea mays*), runner beans (*Phaseolus coccineus*) and courgettes (*Cucurbita pepo*). So what do we mean by 'high temperatures'? Tropical plants will generally not germinate at all below about 12°C, but germination at this temperature is slow and uncertain. To obtain rapid, reliable germination, 20°C or above is required. When does this happen outdoors in Britain? A lot depends on local conditions such as exposure and aspect, but broadly speaking, soil temperatures in southern England are consistently above 12°C from mid-May and above 20°C from late June or early July. With increasing altitude and latitude, everything happens much later. For example, if you garden in Buxton in the Peak District, soils only heat up to 12°C by mid-June, and if you wait for a soil temperature consistently above 20°C, you will be disappointed in most years. In practice, this means that courgettes and runner beans are only worth sowing outdoors in sheltered sites in southern England, and even here you will get an earlier crop by starting them off in a greenhouse or on a windowsill. Everywhere else, germination under glass is essential if you want to start picking before the summer's nearly over.

Since tropical plants need high temperatures for germination, it's only natural to assume that plants from increasingly northern climates need lower and lower temperatures, with Arctic plants needing the lowest temperatures of all. If only life were so simple. As we head north from the equator we eventually arrive in the Mediterranean, and here something very odd happens to germination, for Mediterranean plants generally germinate best at surprisingly low temperatures. For example, *Muscari* species (grape hyacinths), *Chionodoxa* species, *Cyclamen* species and tulips all germinate best between 10°C and 15°C. The strawberry tree (*Arbutus unedo*), a Mediterranean shrub widely grown in gardens, will germinate quite happily at 5°C (i.e. in the fridge). The reason for this apparently odd behaviour is that, from a British perspective, the seasons are back to front in the Mediterranean. Because summers are hot and very dry, much the best time for plants to germinate is in response to the rains of autumn. In order to make sure they are not caught out by a rogue summer shower and then withered by weeks of drought, Mediterranean plants have learned to germinate only at low temperatures, which guarantees that autumn really has arrived.

As we move much further north to the Arctic, you will by now not be surprised to discover that another odd thing happens. Arctic plants, in general, will germinate only at *high* temperatures. Gentians and mountain avens (*Dryas octopetala*) germinate best above 20°C, and the rare Scottish endemic *Primula scotica*, confined to the northern tip of

Scotland, will not germinate at all below this temperature. The explanation, as for Mediterranean plants, lies in minimizing the risks of germinating at the wrong time. In the Arctic these risks are particularly deadly, since winters are extremely severe and, of course, completely dark. Soil temperatures are high in the Arctic during the long days of midsummer and a requirement for these high temperatures means seeds avoid even the smallest risk of germinating in or near the harsh winter.

GARDENERS' TIP

A plant whose germination requirements betray its Mediterranean origins is lettuce. Most varieties of lettuce will not germinate above 24°C (although iceberg types are less sensitive), which can make it difficult to grow lettuce in midsummer. The best way round the problem is to sow in late afternoon and then water with plenty of cold water. Because the temperature-sensitive phase takes place early in germination, this means that with any luck germination will be under way by the following morning. An alternative is to sow lettuce in pots and place them in a shady spot.

For advice on germinating particular plants, try the germination database at www.backyardgardener.com/tm.html.

FIRE AND SMOKE

Viewed from a rain-sodden corner of Europe, it's hard to imagine that over half the world's land surface experiences regular fires. In fact the frequency with which some types of vegetation burn, and the fact that fire is often essential for their very survival, must have come as a real shock to explorers from northern Europe. Not surprisingly, fire has left its mark on seed germination. Many plants of fire-prone habitats accumulate seeds in the soil, and many of these (*Cistus* species are good examples) have hard, impermeable seeds. When you breach the seed coat with a knife or sandpaper, you're doing what fire would normally do in the wild. Seeds respond to fire because much the best time to germinate is just after a fire – there's little competition, plenty of light and nutrients, and most herbivores have been cooked. Rather than shedding their seeds like most plants, some lock them away in tough, woody capsules and release them only after a fire. Familiar examples are the bottlebrushes (*Callistemon* species) from Australia. If you grow these very showy shrubs, you'll have noticed the persistent woody fruits that remain for two or three years after flowering. In the wild the fruits open in response to fire, but in cultivation you simply need to put them in a paper bag and keep them somewhere warm and dry. They will soon open, releasing the tiny seeds. Some plants are not so easily fooled. A once-over with a blowtorch is needed to persuade the massive, cone-like fruits of *Banksia* species (tender Australian shrubs) to loosen their grip on the seeds inside.

For many years, nobody had a clue how to make the seeds of many Australian and South African plants germinate. Botanists from both countries then discovered independently that the answer was smoke. The seeds have evolved to respond to the chemicals in smoke and will now germinate only when they detect these chemicals. Because the discovery was made only in 1990, it's only recently that the active ingredient has been identified. It turns out to be 3-methyl-2*H*-furo[2,3-c]pyran-2-one, a compound no one even knew existed. Not information that will help you in the average pub quiz, but I thought your education was hardly complete without knowing. Also, despite a long list of plants whose seeds respond to smoke, very few of these plants are widely cultivated yet, partly because it was more or less impossible to propagate most of them until recently. Many are also a little tender for cultivation outdoors in the UK.

GARDENERS' TIP

Anyone seriously interested in growing plants from Australia or from the Cape region of South Africa will soon have to acquire an interest in smoke. Fortunately, this doesn't mean lighting a bonfire every time you need to sow some seeds. A need for smoke in a convenient form has created a commercial opportunity that has rapidly been exploited. The South African brand leader, marketed by the National Botanical Institute, is Kirstenbosch Instant Smoke Plus, *reflecting the pioneering role of the botanical garden at Kirstenbosch in the research into smoke and germination. Kirstenbosch smoke is absorbed on to paper and released when the paper is soaked in water. An Australian competitor,* Regen 2000, *is vermiculite impregnated with smoke, which you simply put on top of the compost after sowing the seeds. Ordinary watering then delivers smoke solution to the seeds. For some smoke primer and a starter pack of Protea seeds to try it on, try Trevena Cross nursery in Cornwall (www.trevenacross.co.uk).*

KEEPING SEEDS

Almost as soon as seeds are shed, they begin to age. This ageing takes the form of accumulation of damage to genes and other metabolic machinery. As this damage increases, it becomes manifest first as slower germination and reduced vigour, and ultimately as seed death. Because stored seeds are dry, they are unable to repair any of this damage, and therefore the aim of seed storage is to reduce the rate at which damage occurs. Two factors determine this – temperature and moisture content. As a rule of thumb, the storage life of a seed is doubled by reducing the temperature by 5°C or the seed moisture content by 1 per cent. Storage of seeds for long periods, as carried out in commercial and scientific seed banks, involves drying to very low moisture contents and storage at freezer temperatures. Under such conditions, most seeds can be stored for decades or even centuries. Obviously this is well beyond what gardeners require, but the same principles apply. For storage from one season to the next, no great precautions are required and any cool, dry place will do. For longer periods, it is worth taking the trouble to store seeds in the fridge. Reducing relative humidity by storage in a box with a sachet of silica gel is sometimes recommended, but is really only necessary for storage periods beyond those normally required by gardeners. Increasingly, commercial seeds supplied to gardeners are dried to low moisture contents and sealed in foil packets. Clearly, seeds will store better if such packets remain sealed.

The rules described above apply to almost everything

likely to be grown by British gardeners, but it is worth mentioning those seeds that break all the rules. Seeds of some species (called 'recalcitrant' by botanists) are shed from the parent plant with a high water content and are intolerant of drying. Such seeds are very difficult to store – they normally germinate very quickly and any attempt to dry them results in rapid death. Many tropical crops are recalcitrant, for example oil palm and rubber, and the inability to store their seeds causes a lot of problems for tropical agriculture. Oaks are among the few British recalcitrant species. Acorns cannot be stored and must be sown immediately.

GARDENERS' TIP

Bought seeds can be relied upon to be dry, but great care must be taken to dry self-collected seeds. Quite apart from the direct effect of moisture on seed longevity, seeds are prone to fungal attack above a moisture content of about 12 per cent. Collect seeds in paper bags and spread them out in a thin layer in a cool, dry place until completely dry. Remove all chaff and debris, which will also encourage fungi.

For those who want to take keeping seeds more seriously, the Millennium Seed Bank at Wakehurst Place in Sussex (part of Kew Gardens) will sell you a kit containing all you need for your own 'mini seed bank'. It's £19.95 from Kew or by mail order for £22.95 (tel. 020 8332 5654), and would make an excellent and unusual present for the gardener who has everything.

WHAT ARE F1 HYBRIDS ANYWAY?

Traditionally, cultivars of crops and ornamental plants have been collections of individuals in which each seed or plant is very slightly different from every other. The maintenance of such cultivars requires considerable care to prevent loss of desirable genes or contamination with undesirable genes, and one frequently hears of old cultivars becoming 'weakened' in this way. Often seed companies will offer new 'selections' of old cultivars, from which these undesirable characteristics have been removed.

A relatively recent innovation, first applied commercially to maize, is the production and sale of hybrid F1 seed. The point of such seed is that when two different genetic lines of a species are crossed, the first generation (or F1) resulting from the cross often shows hybrid vigour or 'heterosis'. That is, it is bigger, better and faster-growing than either of the parents. The catch is that to produce this effect reliably, and produce an F1 that is not only better, but *uniformly* so, requires the crossing of two highly uniform, inbred lines. It is the production and maintenance of these inbred lines that makes F1 hybrids so expensive, for a little thought reveals the difficulties involved. Crossing the two parent lines is easiest if they are self-incompatible, that is if one line can only be pollinated by another line. But how do you maintain two pure self-incompatible lines? – because by definition such lines can only be maintained by self-pollination. Brussels sprouts, for example, are strongly self-incompatible and were an early candidate for F1 production. Pure inbred lines of sprouts are maintained by exploiting a loophole in the incompatibility system. The incompatibility gene is only effective in the mature flowers, so pollination is carried out by hand at the bud stage, before the flowers open.

Tomatoes are naturally self-fertile, so maintaining the pure inbred lines is easy. Here the problem is the opposite – preventing self-pollination when the time comes to cross the inbred lines to make the F1 hybrid. This is achieved by removing the male parts of the flowers destined to produce seed at the late bud stage, before the flowers open. Pollination, which usually takes place the next day, is also carried out by hand. This is labour-intensive work and is only economically viable because most of it takes place in India and China, which have the double advantage of cheap labour and a warm climate.

Tomatoes and Brussels sprouts illustrate the difficulties that must be overcome in the production of F1 seeds. Nearly every plant presents different problems and a wide variety of other techniques are available to surmount these obstacles. Some of these techniques are much more complex than those described above, and F1 seeds of many flowers and vegetables have only recently become available to gardeners.

GARDENERS' TIP

F1 varieties are usually claimed to be earlier, higher-yielding and more uniform than traditional flower and vegetable varieties. For commercial growers, uniformity is one of the key advantages, but home gardeners may not be so interested in uniformity and should consider whether F1s are worth the higher price. If you like to save your own seed, remember that the F2 generation will be nothing like the F1 and will usually be much inferior.

Pollinators

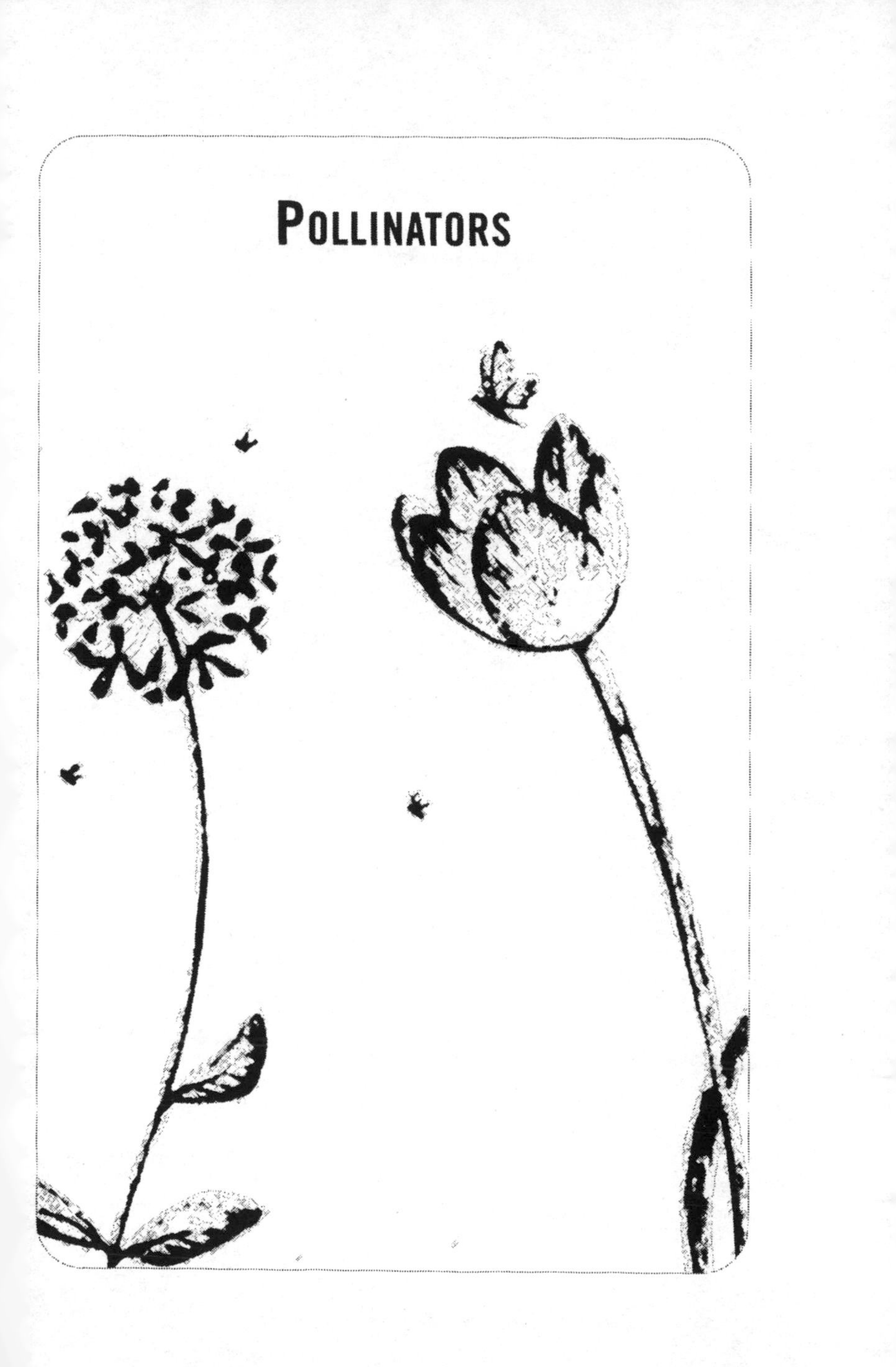

In Praise of Bees . . .

SPARE A THOUGHT FOR POLLINATORS. IN FACT, SPARE several thoughts. Not only are insect pollinators largely responsible for the evolution of the modern plants that make gardening such fun, they continue to work hard on our behalf. The next time you spread honey on your toast, pause to reflect that half a kilo of clover honey represents visits to nearly 9 million flowers, or over 7,000 hours of unpaid bee labour. Honeybees alone added $14 billion to the value of crops in the USA at 2000 prices. Honeybees are in fact rather unusual among bees, in that their colonies ('hives') are perennial, i.e. they survive from one year to the next. Honeybees are only one species of 'social' bee, that is bees that form colonies with a reproductive queen and sterile workers. Honeybees are of course very important pollinators, but not quite as important as some beekeepers would have us believe. The many species of bumblebees, also social bees, are probably as important, and there are in addition very many 'solitary' or 'non-social' bees which are also very good pollinators.

Why are bees such exceptionally good pollinators? The main reason is that they are the only insects in which the whole life cycle is dependent on flowers. Not only do the adults live on pollen and nectar, they also collect these for the larval bees. In all other pollinating insects, pollen or nectar are a food source for the adults only. Thus, of all insects, bees have by far the closest relationships with flowers. The shapes and colours of many flowers are highly adapted to bees. Look at a relatively primitive flower like a buttercup. Like all primitive flowers, it is radially symmetrical and more or less flat. That is, it looks the same from any direction, and its nectar is available to any visitor. Pollination in such a flower is a rather hit and miss affair. More highly adapted flowers tend to be tubular and bilaterally symmetrical, i.e. they have clear left and right sides. Such flowers, for example foxglove, have only one way in, with the stamens and stigma carefully positioned to brush the visitor. Foxgloves are typical bumblebee flowers.

Flower colour is also determined by insect sense organs. Nearly all bees have a similar range of colour vision to humans, but shifted towards the ultra-violet. Thus they can see ultra-violet, which we cannot, but they are red-blind. To a bee, red appears black. So typical bee flowers are usually yellow, blue or purple, and only rarely red. It might seem odd that the common cornfield poppy (*Papaver rhoeas*) is regularly visited by bees, but this flower reflects strongly in the ultra-violet. A further virtue of ultra-violet vision is that soil and leaves reflect very little ultra-violet, so that bees see the world as a rather dull canvas on which UV-reflective flowers stand out.

Decades of work have revealed much about the senses of bees, but it would be a mistake to think we know all there is to know. Part of the problem is appreciating just how different the world looks to insects, which have been following their own distinctive evolutionary pathway almost since the dawn of life on earth. For example, very recent work has shown that bees see only relatively large blocks of colour, while their detailed vision is colour-blind. Perhaps this accounts for the frequent aggregation of small flowers into larger 'superflowers', one of the most consistent and successful trends in the evolution of plants.

GARDENERS' TIP

Like much of Britain's wildlife, many pollinating insects are feeling the pinch these days. Try to grow a wide range of good nectar and pollen sources. This does not *mean just growing native plants or trying to make your garden look like a nature reserve. Pay particular attention to the ends of the season. Willows and ivy are excellent food sources for pollinators active in early spring and autumn respectively. Also, don't forget about nesting sites. Bumblebees nest in holes in the ground or in the bases of grass tussocks, especially on warm, sunny banks. So-called masonry bees nest in house walls, but don't worry, the house won't fall down. And* please *try to tolerate the activities of leaf-cutter bees, which are particularly fond of garden roses.*

TAKE YOUR PARTNERS

For plants as well as animals, inbreeding is usually a bad idea, so plants are faced with a bit of a catch-22. They want to attract pollinators, but they then want them to go away, taking their pollen with them. Not surprisingly, plants have come up with a variety of strategies to make sure this happens. The most foolproof is to do what most animals do: have separate sexes. Red campion, nettle and holly are familiar plants that have gone down this route, so if your holly never has any berries, it may just be a male. A less extreme version of the same tactic is to have both sexes on one plant, but in different flowers. This is a popular option with trees, including birch, hazel and many conifers.

Nevertheless, the great majority of plants have male parts (stamens) and female parts (ovary, style and stigma) together in the same flower. These plants have a whole array of ways of avoiding self-pollination. Often there's just a chemical barrier that prevents pollen from growing on the stigma of the same plant. Sometimes the stigma is receptive only before (or well after) the stamens release their pollen, preventing the pollen fertilizing its own flower.

Some plants are more devious. Most species of *Primula*, including the primrose, *P. vulgaris*, show a condition known in botanical jargon as *heterostyly*. Some primrose plants have a long style, with the receptive stigma at the top of the petal tube and the stamens halfway down the tube. These plants are called 'pin-eyed'. Others, with the stigma halfway down the tube and the stamens at the top, are called 'thrum-eyed'. The two kinds of plants are easily distinguished once you get your eye in. Clearly, pollen that gets on a butterfly proboscis from one kind of plant is in just the right place to pollinate the other kind, but not the same kind, which more or less prevents self-pollination.

Of course, there's no reason to stop at two flower morphs. Some plants have gone further and have flowers with three different style lengths (technically *tristyly*). A good example is purple loosestrife (*Lythrum salicaria*), an attractive native plant for a moist spot or bog garden.

Proving there's no end to the inventiveness of evolution, some plants have come up with 'sideways heterostyly', or *enantiostyly*, in which the style is deflected to either the left or the right. Thus, if you put the two morphs side by side, they look like mirror images. Recent work has shown that the dimorphism is normally controlled by a single gene, ensuring that the two morphs exist in equal numbers. Just like heterostyly, enantiostyly acts as a mechanism to prevent or reduce self-pollination. Bees, which are the typical pollinators of enantiostylous plants, tend to get pollen on one side of their bodies when visiting one morph, which means it's in the right place to pollinate the other morph. Most enantiostylous plants have the morphs on separate plants, but it even works in the odd few plants that have *both* morphs on the same plant. After leaving one flower, even a bee that stays on the same plant has a 50/50 chance of visiting the same morph flower again, which cannot be pollinated by the pollen on the wrong side of the bee. Enantiostyly is rare, but occurs in a range of quite unrelated plants, suggesting that it has evolved on a number of separate occasions. Most enantiostylous plants are tropical, or at least too tender for British gardens, but you don't have to go far to see one. All species of *Saintpaulia* (African violet) have enantiostylous flowers.

BLOOMS FOR BEES

Gardeners who want to attract bees are frequently advised to avoid highly bred modern cultivars and to grow traditional cottage garden plants. But how good is this advice, and is it always right? More importantly, what does this rather vague advice actually mean in practice? Are all modern cultivars bad, and how do you tell? Only recently has anyone tried to answer these questions.

Part of the problem is that plants may be modified in a variety of ways by plant breeders. 'Double' flowers have extra petals, sometimes losing other floral parts such as stamens in the process. In some flowers one or more petals have a spur that projects backwards from the flower. The spur is usually the site of nectar secretion, so spurless cultivars have little or no nectar. Many flowers have bilateral symmetry, with distinct left and right sides. Often, especially in the snapdragon family, the lower lip of the flower closes the entrance to the flower, allowing access only to large, strong insects. In 'peloric' mutants, the flower is less bilateral and also more open. In snapdragons (*Antirrhinum majus*), for example, such cultivars are often called 'azalea-flowered'. Finally, breeders may just make plants bigger – pansies are derived from violas and are basically similar, except that pansies are much larger.

Comparison of cultivars that are close to the ancestral species with those that have been variously modified by plant breeders reveals that most of these modifications cause problems for pollinators, although the picture is complex. Single and double French marigolds (*Tagetes patula*) don't differ much either in rates of nectar secretion or in insect visits. Double larkspurs (*Consolida* species), on the other hand, are useless to bees. Although they have spurs, they secrete no nectar, while the extra floral parts prevent bees from getting at the pollen. Spurless cultivars of nasturtium (*Tropaeolum majus*) have no nectar but are not completely useless, since they are still visited by pollen-collecting bumblebees and honeybees. Surprisingly, despite being bigger all round than violas, and also secreting much more nectar, pansies are generally avoided by insects. Bumblebees visiting violas grasp the base of the lower petal while probing the flower, but seem unable to maintain their grip on the much wider petals of pansies.

Sometimes a modern cultivar may seem to be just as good for pollinators, but careful examination reveals more subtle problems. Normally, snapdragons are the exclusive preserve of large, long-tongued bumblebees, but the peloric form is much more accessible to smaller insects with shorter tongues, especially honeybees. The problem here is that modern intensive farming has not been kind to bumblebees, and the biggest losers have been the longer-tongued species. Flowers that are accessible only to long-tongued bees are therefore particularly valuable in providing nectar that is not available to short-tongued competitors such as honeybees.

GARDENERS' TIP

Flowers are the product of millions of years of plants trying to get the best possible benefit out of their pollinators. Horticultural modification sometimes renders flowers completely useless to pollinating insects, but often has more subtle effects that are difficult to predict. The only safe advice for gardeners who would like to help our native bumblebees is to avoid flowers that show too much evidence of the plant breeder's art.

BUTTERFLIES

Bees use scent and vision in foraging, but many typical bee flowers are not strongly scented to the human nose. Bee-scents are also rather variable, ranging from honey-like scents in sweet violet to the very strong coconut smell of gorse. Lepidoptera (butterflies and moths) seem to prefer heavy, sweet scents; hyacinth and lilac, for example, are pollinated by Lepidoptera. All Lepidoptera have long 'tongues', and typical butterfly flowers have a long tube, with nectar that can only be reached by an insect with such equipment. Red valerian (*Centranthus ruber*) and *Buddleja davidii* are familiar examples. Some of the most heavily scented flowers are pollinated by night-flying moths. Honeysuckle (*Lonicera periclymenum*) shows all the features of a typical moth-flower: a long tube, pale colour and strong scent produced at night. The very common silver-Y moth is one of the most important pollinators in British gardens.

Before leaving insect pollinators, it is worth noting that

bees, butterflies and moths are by no means the only important insect pollinators. In fact almost every flying insect, and many that don't fly, have been recorded as pollinators at some time or other. Probably the most important other group is the very varied true (i.e. two-winged) flies. Some, such as hoverflies, are important pollinators, and visit very much the same sorts of flowers as bees. Many miscellaneous flies, including houseflies and blowflies, also visit flowers. Many of these flies lay eggs in rather unsavoury places, and the scents of flowers that attract flies tend to betray this. Thus the scents of, for example, meadowsweet (*Filipendula ulmaria*), hawthorn (*Crataegus monogyna*), cow parsley (*Anthriscus sylvestris*) and rowan (*Sorbus aucuparia*), while sweet, are not altogether pleasant to many human noses.

GARDENERS' TIP

Many hoverflies are superb bee or wasp mimics, and gardeners frequently have trouble telling the difference. It is an important practical point, since many bees and wasps can sting, but all hoverflies, however fearsome their appearance, are quite harmless. An entomologist will tell you that bees and wasps have four wings, while all flies have only two, which is true but of little practical use. Much more useful is that bees and wasps (and their relatives, the ants) have long antennae. Flies always have short antennae. Oh, and by the way, try to be nice to hoverflies. Not only are they valuable pollinators, the larvae of many species are voracious consumers of aphids.

BIRDS

To the British gardener, birds are pretty unfamiliar pollinators. In fact there are no bird-pollinated flowers native to Europe. Nevertheless, bird-pollinated flowers are not uncommon in other parts of the world, and many have found their way into British gardens. A good thing too, because bird flowers add a whole new dimension to flower form and, especially, colour. Bird vision is particularly acute in the red part of the spectrum, a colour that many insects cannot see at all. Thus typical bird flowers are red, sometimes with a dash of yellow or orange. They also tend to be relatively large, tubular and produce a lot of nectar (it takes a lot of sugar to keep a bird going).

Numerous sorts of birds pollinate flowers throughout the warmer parts of the world, but the most specialized and familiar are hummingbirds, which are exclusively a New World family. There is some truth in the widespread belief that hummingbirds are entirely tropical, but particularly in North America they are migratory, which allows plants pollinated by hummingbirds to grow as far north as Alaska. It also allows British gardeners to grow a wide range of hummingbird flowers, although many are a little tender. Familiar examples are fuchsias, some penstemons, trumpet creeper (*Campsis radicans*) and cardinal flower (*Lobelia cardinalis*). Quite often closely related plants are adapted to different pollinators, and one good example is the large honeysuckle genus. Several species, including the native British *Lonicera periclymenum*, are pollinated by moths, but the orange-scarlet flowers of the beautiful North American trumpet honeysuckle (*L. sempervirens*) are pollinated by hummingbirds. The trumpet honeysuckle reveals the most disappointing thing about bird flowers, which is that they have no scent. Most birds, sadly, have little or no sense of smell.

Before leaving vertebrate pollinators, it is worth mentioning bats. Bat pollination is not infrequent in the tropics, but all bat-pollinated plants are tender and none can be grown outdoors in Britain all year round. The most familiar example in British gardens is the cup-and-saucer vine (*Cobaea scandens*), a perennial grown as a half-hardy annual. It is not all that easy to grow, especially in the north, but worth a try. Unlike birds, bats have a good sense of smell, so most bat flowers have quite a strong scent, not much like that of insect-pollinated flowers. Have a sniff at *Cobaea*. See what you think.

GARDENERS' TIP

It is a pity that European gardens lack bird pollinators. One of the chief joys of gardening throughout the warmer parts of North and South America is hummingbird watching. But wildlife gardeners should definitely go ahead and grow as many bird flowers as they want – they are terrific nectar sources for insect pollinators.

Gardens and Wildlife

The Wild Garden

IN 1971, JENNIFER OWEN, FORMERLY A PROFESSIONAL biologist, but now retired, set about systematically recording the wildlife of her rather ordinary garden in the Leicester suburb of Humberstone. In 1991 she published a book, *The Ecology of a Garden: The First Fifteen Years*, that should be required reading for gardeners everywhere. Its first chapter contains a perceptive paragraph that is worth quoting in full:

> *There is a fashion dating from about 1980 for 'wildlife gardening', although this is usually interpreted as emulating the countryside. It is far from being a case of laissez faire, for the wildlife gardener puts a lot of time and labour into creating wildflower meadows, marshy pools, or even woodland walks. Few people attempt to improve, as wildlife habitats, ordinary gardens that are productive and attractive in the conventional sense; instead, an effort is made to create outside the backdoor a most ungardenlike habitat. This is admirable in its own way, but overlooks the possibilities of having both the productive garden that most people want, and a wildlife habitat at the same time.*

The extent of these possibilities is amply demonstrated by Owen's book. For fifteen years she recorded all the plants and animals in her garden. We don't need to worry here about how she did this, but it is worth noting at the outset that she made a very thorough study of some groups of insects and paid little or no attention to others. It is also worth noting that she gardened 'organically' only in the sense that she used no pesticides of any sort. She did, however, use chemical fertilizers (Growmore). Below I briefly summarize her findings under the headings of the various groups of plants and insects. I consider vertebrates, including birds, later.

In 1984, a typical year, the garden contained 264 species of flowering plants in an area of 741 square metres. Some were planted, some arrived on their own. This is a level of plant diversity not matched by any natural vegetation anywhere in the world, not even tropical rainforest. More about the relative merits of native and cultivated plants later.

A total of twenty-one species of butterflies (about one third of the British list) were seen in the garden, but many only rarely and some only once. Only the whites (small white, large white, green-veined white and orange tip) actually bred in the garden, the first two on brassicas, the second two on other members of the cabbage family, including *Aubrieta*, *Arabis*, candytuft (*Iberis*), dame's violet (*Hesperis*) and honesty (*Lunaria*). The big, attractive members of the Nymphalid family (peacock, small tortoiseshell, red admiral and painted lady) were common but did not breed, although their larval food plants (thistles or nettles) grew in the garden.

Britain has very many more moths than butterflies and Owen recorded 263 species. Many of these bred in the garden, and the plant species they used shed interesting light on the relative values of native and alien plants. Of the most frequently used larval food plants (i.e. those used by at least five species of moth), six were native and nine were alien. The single most popular species, employed as a larval food plant by eighteen moth species, was *Buddleja davidii*. Not only is *Buddleja* an alien, the plant family to which it belongs does not occur in Europe.

These moth data are worth a closer examination, because it has become an article of faith among wildlife gardeners that 'native is best'. This belief can be traced back to work on trees which showed that natives such as oak and birch have many more insect species eating them than foreign trees such as horse chestnut and sycamore. People have tended to assume, with little or no evidence, that this result can also be applied to shrubs and herbaceous plants. Unfortunately, entomologists sometimes forget just how poor the data on insect herbivores really are. The shrubby cinquefoil, *Potentilla fruticosa*, is a very rare native shrub that is very widely grown in gardens. No fewer than nine moth species used *P. fruticosa* as a food plant in Owen's garden, although in the wild no moths are recorded as eating it, which shows just how easily rare plants are overlooked both by insects and by entomologists. Some insects are also remarkably catholic in their tastes. Caterpillars of the pretty angle shades moth ate thirty-one different plants in Owen's garden, twenty-five of them alien garden plants.

A total of 133 species of hoverfly have been recorded in Leicestershire, and Owen recorded ninety-one of these. Nearly all the common garden hoverflies have larvae that eat aphids. Nearly 100 species of bees and wasps were found in the garden. Although most of the social bees (bumblebees) were recorded, none bred in the garden. Owen also found over a third of the British species of ladybird, although *the* ladybird (the seven-spot) was not the most abundant. The commonest ladybird, and the one that bred most often in the garden, was the smaller two-spot. Like its larger cousin, its larvae are great consumers of aphids.

Owen eventually found 2,204 species of animals and plants in her garden. Making some reasonable assumptions about species that were missed and whole groups that were not studied at all, the garden may have contained 8,500 species of insects alone. Why so many? Part of the answer is the extraordinary diversity of plants and habitats found in the average garden. More plants means more herbivores, which in turn means more predators and parasites. Gardens also contain the extremes of structural diversity, from bare ground through shrubs to mature trees. Ecologists have long known that 'edge' habitats (e.g. woodland edges) have a high diversity of wildlife, since they accommodate species of the neighbouring habitats and those that prefer the unique conditions of the edge itself. Gardens are nearly all 'edge' of one sort or another: where lawn meets herbaceous border, pond meets lawn and shrubbery meets compost heap. In the majority of insect species, juveniles (larvae) need different conditions from the adults, and these different conditions are more likely to be adjacent in a garden than anywhere else. In short, if you had to design a nature reserve to pack as much biodiversity as possible into a small space, the result would probably look a bit like a slightly untidy garden.

GARDENERS' TIP

There is already a lot more wildlife in your garden than you are aware of, and you can encourage more without overt 'wildlife gardening'. There are few rules, but avoid pesticides if you can, and don't be too tidy. A wide variety of plants and habitats is a good idea, and trees and large shrubs are particularly valuable, but don't think that you have to do everything in your *garden. The latest research shows that most insects are highly mobile and are certainly not confined to single gardens. They may, for example, breed in one garden but forage for food in several. Perhaps the best thing you can do for local wildlife is try to provide something lacking from surrounding gardens.*

Leave clearing up and pruning until late winter whenever possible. Many insects overwinter in the stems and seed heads of herbaceous perennials. Grow lots of different plants and keep the soil well covered. Don't waste time trying to provide larval food plants for the high-profile Nymphalid butterflies. There are enough nettles and thistles without you growing any more, and the chances are they won't like your *nettles anyway. Grow plants with big flat flower heads, like fennel* (Foeniculum vulgare) *and* Achillea, *to attract hoverflies. Attract enough hoverflies and they will solve your aphid problems for you.*

WHY ARE THERE SO MANY INSECTS?

Why are there so many different kinds of insects in Jennifer Owen's garden, and what on earth are they all doing? I don't think anyone could give you a definitive answer to either of these questions, but here are a couple of things worth thinking about.

First, size. It is one of the few 'rules' of ecology that there are more small animals than large ones. Consider how many individual mammals the average garden might support. Maybe a family or two of mice, a couple of squirrels and hedgehogs, perhaps a part share in a fox – say a dozen or fifteen individual animals in all. Given these low numbers, it's not really possible to have many different *kinds* of mammals in a typical garden. It's therefore not surprising that a recent survey found that just five mammals are encountered in the average garden – mice, grey squirrels, hedgehogs, foxes and bats, probably nearly always pipistrelles. These are the only mammals I see in my garden, and the only ones in Owen's garden too, except for a dead vole dropped by a passing kestrel. Much the same applies to birds. There are about 125 million individual breeding* birds in Britain, which sounds a lot until you consider that's just two birds per person. So if you have a pair of blackbirds or blue tits nesting in your garden, take good care of them – they are *your share* of Britain's birds.

On the other hand, how many individual insects are there in your garden? Frankly, I haven't a clue and neither has anyone else, but consider that a hectare (that's a 100 metre square) of ordinary countryside may contain over 5,000 million aphids alone. The potential to have many dif-

*Populations of birds fluctuate violently, and there are many more than this in mid to late summer when the numbers are swollen by juveniles, most of which do not survive their first winter.

ferent *species* of aphids, beetles, flies and everything else is clearly enormous. Small size opens up other possibilities. A garden offers so many more nooks and crannies to small animals than to large ones. To take just one example, only really small animals could spend part of their life cycle *inside* a leaf, yet this is one of the most successful and abundant insect lifestyles. There may be up to 500 species of leaf miners in Britain – mainly flies, but also beetles and moths. Extracting the ubiquitous holly leaf miner from its refuge is popularly supposed to have given blue tits the idea for tearing the tops off milk bottles.

A second factor that contributes to the huge diversity of insects is that, quite apart from small size, insects have discovered ways of making a living that simply don't exist among vertebrates. A typical vertebrate food chain has just two components – herbivores eaten by carnivores. Sometimes a top carnivore adds an extra link, but that's about it. To understand how insects have gone one better, you need first to appreciate the distinction between a predator and a parasite. Predators kill their prey (obviously) and a single predator usually kills several prey during its life. Parasites, on the other hand, rarely kill their hosts, and a single host may support many parasites. Many insects, usually called *parasitoids*, have adopted a lifestyle that combines elements of both. Parasitoids usually develop inside their hosts (like a parasite), but the host is always killed (like a predator). Adult parasitoids search for their prey, usually the eggs, larvae or pupae of other insects, then lay one or more eggs on, or more often in, the host. The young parasitoid develops inside the host, consuming it entirely apart from the skin, then pupates and eventually emerges as a new adult. In other words, the monster from *Alien* is alive and well and living in your garden. Most parasitoids are wasps; some are large insects, and those that attack hosts hidden deep inside plant stems may have fearsome egg-laying ovipositors, but most are small and rather inconspicuous. Indeed, the smallest of all insects are parasitoids. Naturally, potential hosts are not keen on being found, and there is thus a constant state of warfare between parasitoids and their hosts, with intense pressure on parasitoids to use all available means of host location, and equally intense pressure on hosts to be as invisible as possible. Even if hosts get very good at this, they can still be outsmarted. On the principle that 'my enemy's enemy is my friend', some plants produce chemicals that attract the parasitoids of the aphids that attack them. Further up the food chain, aphids may attract the parasitoids that attack the hoverfly larvae that eat *them*.

All this would be interesting enough, but the really remarkable thing about parasitoids is their almost unbelievable diversity. No one is sure how many there are, but the current best guess is around 1 million species worldwide. In one study of the insects living on a single plant species (broom) at a single site in southern England, there were more species of parasitoids than all the herbivores and predators combined. But this sort of information is rare, because even most professional entomologists are unable to recognize more than a few of the more conspicuous kinds. Jennifer Owen was fortunate to be able to make a particular study of one family of parasitoid wasps, the Ichneumonidae, although there are several others. The diversity of ichneumonids in the Leicester garden was astounding. Owen recorded 533 species, of which fifteen were new records for Britain and four were completely new to science. Clearly you don't have to go halfway up the Amazon, or indeed anywhere, to discover a new species.

GARDENERS' TIP

Parasitoids are of major importance as biological control agents, worth millions of pounds every year. One of the first to be commercially introduced, and still one of the most successful, is Encarsia formosa, *a parasitic wasp used to control greenhouse whitefly. There are now biological controls for most greenhouse pests, and many of them are parasitoids.*

Don't think, however, that you only get help from parasitoids you buy. They are working in your garden every day, mostly unseen, and most outbreaks of aphids or caterpillars are stopped in their tracks by parasitoids before you even notice them. Although more attention is paid to high-profile predators like ladybirds or lacewings, parasitoids play a key part in the functioning of a healthy garden.

SLUGS HAVE FEELINGS TOO

There is a contradiction at the heart of wildlife gardening which, like some awful relative you hope will never visit, is never mentioned but always lurking at the back of your mind. This is that there is *good wildlife* and *bad wildlife*. In order to try to illustrate what I mean, let's look at a recent initiative by the Countryside Council for Wales (CCW), one of the three British government national conservation agencies. In an entirely admirable attempt to end some of the confusion about which plants should be grown by wildlife gardeners, they have attempted to produce a definitive list of wildlife-friendly garden plants (see the full list on the CCW website at www.ccw.gov.uk). It's a good list, and the wildlife in your garden would undoubtedly benefit if you grew a selection of the plants on it. The really interesting thing, however, is why these 174 plants are recommended. Because some plants are recommended for more than one reason, the list actually contains a total of 276 different 'plant/reason' combinations. Forty-six of these concern birds and are nearly all plants with berries. Virtually all the remaining recommendations (211 out of 230) involve flowers for pollinators, in the following order: bees, butterflies, other insects, hoverflies, moths. This list therefore reflects two things – first, that gardeners like pollinators, especially bees and butterflies, and second, that the ecology of pollination is reasonably well understood.

So far so good, but consider hoverflies – large, harmless, attractive, often brightly coloured insects, and good pollinators. Just the sort of thing you want in your garden. Oh, and even better, the larvae of many species eat aphids. But there's a snag here: in providing lots of flowers, you are catering only for *adult* hoverflies. Hoverfly larvae need lots of aphids, so shouldn't you be providing them too? In fact aphids are the garden equivalent of krill – small, extremely abundant, and the base of a whole ecosystem of predators and parasitoids. To take another example, there are fourteen recommendations in the CCW list of plants for moths, eight of them for flowers and six as larval food plants. *Six?* There are 886 species of large moths in Britain, and Jennifer Owen recorded 263 of them in her garden. The single most popular larval food plant, as I've already mentioned, was *Buddleja*, which is on the CCW list, but only as a source of nectar for butterflies. Of the six larval food plants for moths on the CCW list, only one is in the top fifteen moth food plants in Owen's garden.

To summarize the problem: first, at the base of all food chains are animals that eat plants, but few gardeners would willingly grow plants that could be relied on to have half their leaves devoured by caterpillars, with the remainder infested with aphids. On the contrary, if we knew a plant that would kill stone dead any aphid or caterpillar that touched it, most of us would probably grow it without a moment's regret. Second, the ecology of garden plants is generally so poorly understood that even if you wanted to grow plants that were attacked by a wide range of herbivores, you probably wouldn't know where to start.

The real paradox in recommending pollinator-friendly garden plants is that with a few exceptions (some modern cultivars, and wind-pollinated plants like grasses), *all* garden plants are at least moderately useful to pollinating insects. One symptom of this is that while the CCW list is one of the best of its kind, I'm sure we all have favourite bee or butterfly plants in our own gardens that are not on it. One example at random from my own garden: sage (*Salvia officinalis*). And although I fully understand why *Buddleja* continues to appear on lists of plants for butterflies, can there be a single gardener left with even the slightest interest in wildlife who doesn't already grow it?

GARDENERS' TIP

Before you embark on wildlife gardening, some serious self-examination is in order. Growing the plants on the CCW list will do for starters, but real wildlife gardening is less about what you grow than about how you grow it. You'll know you're a real wildlife gardener when you can agree with Jennifer Owen, even if through gritted teeth: 'There are no pests, because everything in my garden is a source of interest and enjoyment.'

BITS OF COUNTRYSIDE, OR WHY IS WILDLIFE GARDENING SO DIFFICULT?

Even after being alerted to the potential for wildlife of ordinary gardening, you may still want to try some conventional wildlife gardening. For some reason, this almost always involves trying to create a wildlife meadow. To dispose of one obvious problem without delay, remember that meadow plants are all herbaceous perennials. Pictures of wildflower meadows in books and seed catalogues often seem to include annuals like poppies and cornflowers, but these will not survive in a meadow. However, you *can* grow them in a wildlife garden – see later.

Before trying to create your own patch of flowery grassland, it is worth spending a minute looking at real, wild grasslands. A glance at the average grassy field soon reveals that it often contains just that – grass, sometimes with the odd dandelion or hogweed. In other words, not what we want. Fields of interesting grass, with lots of different plants and flowers, would have been familiar enough to your grandparents but are distinctly rare these days. The key to this loss of diversity (and thus the key to creating a wildlife meadow) is nutrients, which provides an excellent demonstration that you *can* have too much of a good thing. A recent survey across Europe looked at the number of plant species in patches of grassland, each 100 square metres, and related this to soil nutrient content. The results were clear – although many grasslands contained between twenty and forty different species, and the most diverse contained sixty, no grassland with a phosphorus level above 5 milligrams per 100 grams of soil contained more than twenty species. Even worse than that, the plants on high-phosphorus soils were always the wrong ones, mainly coarse grasses. Rare plants, which are often those with the prettiest flowers, were just missing completely above 5 milligrams of phosphorus.

So why is phosphorus a problem? First because it is one of the 'big three' nutrients (NPK – nitrogen, phosphorus, potassium) that large, coarse plants need lots of to grow big and fast, thus excluding smaller, slower-growing species. Second, nitrogen and potassium are dynamic, here-today-gone-tomorrow sort of nutrients, but phosphorus is much more persistent. A soil with lots of phosphorus is likely to stay that way. In fact recent research in both France and Belgium has shown that soil cultivated by the Romans (but not since) is still clearly recognizable from its elevated phosphorous content. Third, most fertile agricultural and garden soils contain much more than 5 milligrams of phosphorus, often somewhere between 10 and 40 milligrams.

In practice this means that creating a species-rich wildflower meadow on a typical garden soil is about as easy as drinking tea with a fork. The coarse grasses in such a meadow will always be trying to strangle the plants you want to grow. Since too many nutrients is a problem for those concerned with nature conservation as well as for gardeners, you may wonder if science has come up with any way of reducing the nutrients in soil. Unfortunately, the answer is not really, although not for lack of trying. Cutting the vegetation and removing the cuttings (in other words, traditional hay-meadow management) works, but only very slowly. Removing the top soil and starting again on the nutrient-poor subsoil certainly works, but is expensive and a little drastic for gardens. Inverting the top 30–60 centimetres of soil to bring the subsoil to the surface is easier, but only just. Attempts to 'immobilize' the phosphorus as insoluble aluminium phosphate by adding large quantities of aluminium sulphate are promising, but are still at an experimental stage and cannot yet be recommended to gardeners. You could also sow yellow rattle (*Rhinanthus minor*) in your meadow. Yellow rattle is a *hemiparasite*, which means it has chlorophyll (unlike a complete parasite), but attaches itself to the roots of other plants, normally those of the most vigorous plants around. Not surprisingly, this seriously reduces their vigour, which has very much the same effect as lowering the fertility. Yellow rattle cannot be guaranteed to succeed, but it's worth a try.

One final option is to add some carbon. Soil is full of micro-organisms that compete for the same nutrients as plants, so if you can give them an edge in that competition, you are effectively lowering the fertility for the plants. Since microbes are always hungry for carbon, one way to do this is to add some nutrient-free carbon. Untreated wood shavings or sawdust are good, especially if there's a local sawmill or timber yard that will give you some for nothing. The problem is that to be really effective, both need digging into the soil. A lazy alternative is to water your meadow with sugar solution, which may sound bizarre (and will certainly have to be kept a secret from your neighbours), but it does work.

Even if you can't produce a flowery meadow, however, leaving even a boring patch of long grass can have major benefits for wildlife. Long grass may encourage grass-feeding butterflies to breed or bumblebees to nest and will provide a habitat for some insects, such as grasshoppers, that is unlikely to be found anywhere else in the garden. If leaving all or part of your lawn uncut offends the neighbours, try sticking a large sign in the middle, saying: 'I do own a lawnmower – I am only doing this to encourage wildlife.'

GARDENERS' TIP

If you have any choice in the matter, start your meadow in the least fertile part of your garden. Avoid anywhere that has received lots of fertilizer, such as an old vegetable plot. The most certain way to establish a meadow is to remove the existing vegetation and start again, but you may not want to go this far. If starting from scratch, do not sow grasses (they will turn up anyway), and sow plenty of flower seed.

Most gardeners will probably prefer to try to establish meadow plants in some existing grass. Using plants rather than seed is the most reliable way to do this. You are frequently advised to grow large plants and then cut a hole at least 30 centimetres across to plant them in. In fact the benefits of doing this are debatable, since whatever you do, your flowers will always struggle if your meadow is too fertile. A better bet is to make sure you grow robust meadow plants such as Malva moschata *(musk mallow),* Persicaria bistorta *(bistort),* Centaurea nigra *(knapweed),* Ranunculus acris *(meadow buttercup),* Primula veris *(cowslip),* Knautia arvensis *(field scabious),* Rumex acetosa *(sorrel),* Leucanthemum vulgare *(ox-eye daisy),* Lathyrus pratensis *(meadow vetchling) or* Geranium pratense *(meadow cranesbill). Have a good look at the plants in local road verges, since they will at least be suitable for your soil and climate. You can also collect seeds of common species from road verges – technically you should ask the landowner for permission, but in reality your local highway authority is unlikely to object. Spring may seem like the obvious time to plant, but in fact autumn will do just as well and may even be better. Only purists will complain if you also include some foreign meadow plants, such as* Papaver orientale, Hemerocallis *species,* Lupinus polyphyllus *or* Lychnis chalcedonica.

You can start your meadow plants from seed, but it is a waste of time just sowing into intact turf. Cut gaps at least 30 centimetres in diameter and sow plenty of seeds – only a very small proportion will establish. Sowing on top of a layer of sharp sand (at least 2.5 centimetres, preferably 5) will prevent most buried weed seeds from germinating.

Cut your meadow at least once a year, remove the cuttings, and no fertilizer.

So far we've considered only the survival of your sown plants, but of course you would really like these plants to spread by seed into the surrounding grass. This means you must let your plants seed, which in turn means not cutting before the end of July at the earliest. If you grow late-flowering species such as Succisa pratensis *(devil's bit scabious), you will have to cut as late as September, even if only occasionally. Cutting late (not before the end of August) has other benefits. If meadow brown butterflies like your meadow, they should all have emerged by then. On the other hand, if you or anyone else in your family suffers from hay fever, you may feel compelled to cut the grass soon after it starts flowering. I'm afraid that's what happens in my garden. Recent research suggests that a good time to cut, if you want to control coarse grasses and favour wild flowers, is midwinter. January is ideal.*

ANNUAL MEADOWS

An easier alternative to your own meadow is your own arable field. Many popular hardy annuals are former arable weeds. The key to their decline is mainly the increasing use of herbicides, but here also high nutrient levels have played a part. Many of our rarest arable weeds used to grow on relatively poor soils. Arable weeds are all annuals and depend on a major annual disturbance, provided naturally by ploughing. Ploughing prevents the establishment of perennials and provides ideal conditions for the crop and weeds to establish every year from seed. In a garden, this is much easier to arrange than a meadow, since it is actually rather similar to the way hardy annuals would normally be grown, although you are relying on self-seeding to keep things going. A collection of arable weeds will also put up with more fertile soil than a real meadow, although even here, too many nutrients will encourage the wrong sort of plants to take over.

GARDENERS' TIP

Choose a sunny, open spot. Thoroughly cultivate the area to remove any perennial weeds and sow a mixture of hardy annuals in spring. Just as when sowing meadow plants, applying a layer of sand and sowing your seeds on top will prevent most buried weed seeds from germinating. Good plants to try are 'honorary natives' (in fact ancient introductions) such as Papaver rhoeas *(poppy),* Centaurea cyanus *(cornflower) and* Chrysanthemum segetum *(corn marigold), and aliens such as* Eschscholzia californica *(Californian poppy),* Rudbeckia hirta *(black-eyed Susan),* Papaver somniferum *(opium poppy) and* Atriplex hortensis *'Rubra'. Thereafter just rake off the dead plants in the winter and thoroughly disturb the soil, adding a few more seeds of anything that looks like dying out. To make your 'field' less dull in the spring, plant some bulbs. As a variation, add some biennials, either natives such as foxgloves and teasels or aliens like* Hesperis matronalis *and* Salvia sclarea *var.* turkestanica, *and cultivate only every two years. For annual and perennial seed mixes developed after extensive trials at Sheffield University, try www.pictorialmeadows.co.uk.*

And again, never add fertilizer of any sort.

THINGS WITH FOUR LEGS . . .

Some four-legged wildlife is just a menace (e.g. grey squirrels and rabbits). Others are independent souls and are difficult to encourage or discourage (e.g. foxes), while some sound like a good idea until you actually experience them (e.g. badgers). Much four-legged wildlife is abundant but inconspicuous (e.g. mice and voles). There are – trust me – far more mice in your garden than you think there are.

Mammals, like all other animals, become more common as they get smaller, and they don't come much smaller than mice. Surprisingly, however, if you compare British birds and mammals of about the same size, mammals are about *forty-five times* more abundant. Gardeners can be forgiven for greeting this fact with scepticism; after all, gardens are clearly full of blue tits and blackbirds, but not noticeably awash with mice, voles and shrews. Nevertheless, the data on British birds and mammals are the best in the world, and undoubtedly true.

Which raises the obvious question: why do garden birds appear to be more common than similar-sized mammals when they are in fact much rarer? The answer is that most birds are conspicuous, noisy and active during daylight, while most mammals take care to remain hidden and (most importantly) nearly all are nocturnal. Consider how often you see squirrels, which are active by day, and how rarely you see hedgehogs, which are not. As to *why* birds are so much less abundant than similar-sized mammals, I'm afraid that's one of those awkward questions no one has figured out yet. Whatever the reason, it's certainly true that there is probably more wildlife in your garden than meets the eye.

There isn't a huge amount you can do to either encourage or discourage mice. But some animals can be encouraged and will add significantly to the enjoyment of your garden. Perhaps top of the list here are hedgehogs and amphibians. Hedgehogs are common in gardens throughout mainland Britain. They like anywhere there is plenty to eat, and this includes gardens, especially suburban ones. They eat beetles, worms, caterpillars, slugs and almost anything they can catch, but little plant material. They will take eggs and chicks of ground-nesting birds, but foxes, cats, crows and magpies are much worse in this respect. They carry fleas, though not the same sort as cats or dogs. Females have litters of four or five young (sometimes more), between April and September, and the young need to weigh at least 450g by November or they are not fat enough to last the winter. Hibernation usually begins about November and ends around Easter, but depends a lot on the weather. The winter nest is made of leaves, tucked under a bush or log pile or garden shed, anywhere that offers support and protection.

Before leaving hedgehogs, one final thought. Hedgehog populations may be five or ten times denser in gardens than in the countryside, and one reason for this is predation by badgers. Badgers will eat hedgehogs if given the chance, and are strong and dextrous enough to prise open a curled-up hedgehog. Because badgers don't like towns, urban gardens provide space for hedgehogs that is largely free of their main predator. Neither cats nor foxes can cope with adult hedgehogs, but large dogs can, and dogs may attack hedgehogs if they come upon them sleeping in their daytime nests. Keeping your dog under control at all times will reduce the chance of your local hedgehog ending up as pet food.

Perhaps the most obvious vertebrate wildlife in gardens is the amphibians attracted by garden ponds. Frogs and toads are most likely to colonize your garden pond, although frogs are both more abundant and generally prefer smaller ponds, so are most likely to prefer modern small gardens. Both are omnivorous as tadpoles but completely carnivorous as adults. Slugs and snails are high on their list of favourite foods, so they are both extremely useful animals to have in the garden. All three British species of newt are in decline and are now distinctly rare in gardens. Unlike frogs and toads, they are carnivorous throughout their lives, and are also great consumers of slugs and snails as adults.

Britain has few reptiles and the only one likely to be frequently encountered in the garden is the slow worm. Although it looks like a slightly squared-off snake, the slow worm is in fact a legless lizard. If you have them in your garden, count yourself extremely lucky – no other animal has quite such a fondness for slugs and snails, or appears to show such obvious relish while eating them.

Finally, moles. Most gardeners are a bit ambivalent about moles. They are undoubtedly cute, and we all remember the amiable Mole in *The Wind in the Willows*, but few things drive gardeners into a fury more reliably than molehills on a neatly-trimmed lawn. In practice, there are only two things you need to know about moles. The first is that the various sonic, vibrating or whatever devices that are sold to repel them are no use at all. This is not to say that one day an effective mole-repellent for gardeners will not be invented, just that it hasn't been yet. The second thing is that moles consume prodigious numbers of earthworms, and so it should be possible to control moles by controlling worms. For a long time nobody was sure whether reducing earthworm numbers might actually *increase* mole damage, as the moles dashed hither and thither in an increasingly frantic search for fewer worms. However, recent work conclusively demonstrates that reducing worm numbers also reduces the number of molehills. Since abundant earthworms are one sign of a healthy garden, this might not seem very useful information, but you only need to reduce their numbers under your lawn. The easiest way to achieve this is to increase the acidity of your lawn by frequent applications of ammonium fertilizers, which will have the useful side-effect of discouraging most common lawn weeds.

GARDENERS' TIP

Hedgehogs are becoming rarer, although they are not endangered, so gardeners can play a big part in maintaining their population. As in the case of much wildlife, excessive tidiness is the great enemy of hedgehogs. Lack of hibernation sites, rather than lack of food, is probably the main reason for a scarcity of hedgehogs. The basic material of a nest is tree leaves, so do not clear them all up. Hedgehogs hibernate under garden bonfire heaps, so always check before burning. They are not so fussy about nesting sites, but again the long grass and weedy patches they like are missing from tidy gardens. Although hedgehogs swim well, they easily drown in smooth-sided garden ponds. Ponds (and swimming pools) should have a piece of chicken wire dangling into the water to help the animals climb out. Feeding hedgehogs is a good idea. Bread and milk will do no harm as long as they have plenty of other things to eat, but dry (not tinned) dog or cat food is better.

To attract amphibians, the obvious requirement is a pond. This should have at least one gently sloping side so that animals can enter and leave easily, and plenty of cover nearby. Tall vegetation and a pile of stones or logs is ideal. Oh, and of course – no fish. *Slow worms like to hide beneath pieces of wood, flat stones or even corrugated iron, so provision of these hideaways will encourage them to stay if they live in your neighbourhood.*

Slug pellets are dangerous to hedgehogs, amphibians, reptiles and most other garden wildlife.

. . . AND TWO

Wood pigeons decimating winter brassicas notwithstanding, birds are welcome visitors in most gardens. Indeed, many gardeners devote some time and effort to attracting birds. Much of this effort goes on provisioning a bird table or other forms of artificial feeding, and although you may think there is little new to say about this, feeding birds is not completely straightforward. First, use your imagination. Don't just stick to peanuts or sunflower seeds; you could also try scraps such as cooked rice or potatoes, cheese, fat, stale cake, breadcrumbs and fruit. Also, because different birds like to feed at different heights, the really keen bird feeder will put these both on a bird table and on the ground. Second, don't forget fresh water. Make sure it's clean and regularly topped up in summer, and replace with fresh if it freezes in winter. Third, something those who keep chickens know all about, but the rest of us tend to forget: calcium. Birds need lots of calcium to lay eggs with good, strong shells. Poultry farmers buy sacks of ground oyster shells, but you can do the same for nothing – just add some smashed egg or snail shells to your bird table. It's also a good excuse to bash the snails in your garden with a clear conscience.

GARDENERS' TIP

A recent survey reveals that at any one time, 80 per cent of UK bird feeders are empty. If this is because you can't be bothered to go out and buy more food, or because you started to feed the birds and then found it was too expensive, buy food more cheaply in large quantities from bulk suppliers. There are plenty on the internet, but I use www.wildbirdfood.uk.com.

So much for bird feeders, but what about the food provided by the garden itself, in particular, berries? Given limited space, which shrub and tree species will provide the best food for birds, and which will attract the greatest variety? A major consideration is the seasonality of fruit availability. Most shrubs have berries during late summer or autumn, with currants and cherries about the earliest to ripen. Birds, however, or at least residents and winter visitors, need feeding in winter too. Indeed birds need a lot of energy in winter to keep warm, and the very survival of some species is determined largely by the ready availability of winter food. The great mainstays of British winters are holly and ivy, and of these the least reliable is holly. This is because holly ripens relatively early, usually in October, and bushes are often completely stripped of berries by the end of January, or even December. The great majority of observations of birds feeding on holly are in December and January. The survival of berries beyond this period depends on a bush being defended by mistle thrushes. Mistle thrushes are the largest of British thrushes and routinely defend holly bushes against other birds to provide a long-term supply of berries for themselves in late winter. They also defend mistletoe in the same way (the bird's name comes from this habit), but this is of less interest to gardeners, since mistletoe is not common in gardens and is difficult to cultivate. Defended holly has the longest period of availability of any berrying shrub, going almost right around the calendar.

Ivy flowers in autumn and the fruit ripens in winter, or sometimes not until early spring. Ivy thus provides a double bonus to wildlife. Its nectar-rich flowers are the last big source of nectar for insects before the winter and its berries ripen at a time when there is very little other fruit available. Its fruits are also exceptionally nutritious, being unusually rich in fat.

Holly and ivy, although very often grown in gardens, are of course native shrubs. Among the alien garden plants useful to birds, *Cotoneaster* and *Pyracantha* are pre-eminent. There are many species and varieties of both, and all provide valuable winter food for much the same birds as the native shrubs. Bushes of both may sometimes be defended by mistle thrushes, and at times by other members of the thrush family, such as blackbirds. Of course, many other garden shrubs produce fruit, including *Viburnum*, *Mahonia*, *Berberis* and ornamental cherries and hawthorns, but few rival *Cotoneaster* and *Pyracantha* for the consistency and quantity of their fruit crops.

Which birds can you expect to be attracted to your garden by berries? Principally the thrushes (song thrush, mistle thrush, blackbird, fieldfare and redwing), the robin, the starling, three crows (carrion crow, magpie and jay), and maybe a warbler or two (most likely the blackcap). This is not an exhaustive list, but other fruit-eating visitors are all much less likely. If you are lucky, your garden may experience a winter invasion of waxwings from Scandinavia. Of all European birds, waxwings are the most specialized for a fruit diet – their particular favourite is rowan berries. Normally, however, much the largest quantity of berries is likely to be consumed by the thrushes, and pre-eminently the blackbird, Britain's commonest garden bird. This is because fruit, when available, forms a large part of all their diets and because they are relatively large birds, so they can consume most sorts of berries. This is a crucial point – very many native and garden shrubs have berries in the 8–11 millimetre range, and birds need to have beaks big enough to swallow fruit of this size. The smaller fruit eaters, such as the robin and the warblers, tend to concentrate on smaller fruits, such as dogwoods (*Cornus* species), spindle (*Euonymus* species), honeysuckle, elder and privet. Oddly enough, starlings have a similar problem, despite being relatively large birds. This is because their narrow, pointed beaks are well adapted for poking in the ground, but this makes them inefficient fruit eaters. The most popular shrub with small berries, eaten by all the birds in the above list, is elder. Elders have abundant, nutritious fruit, and are available in August and September, before the fruits of many other shrubs ripen. Starlings in particular love elderberries, but they are also much favoured by robins and blackcaps. Note that although we associate berry-eating birds with winter, warblers are all summer visitors, and elders and other early-ripening species are important for fattening up before setting off on the autumn trek to Africa.

Many familiar garden birds never, or only rarely, eat fruit. Sometimes this is not obvious, since some birds pluck berries in order to get at the seeds inside. The specialist seed-feeding finches, particularly the greenfinch, will often do this, as will several species of tits.

GARDENERS' TIP

If you like birds, grow plenty of berrying shrubs. It matters not at all whether these are native or alien – the birds don't mind. As explained above, ivy is particularly useful for birds, but you need patience to persuade ivy to flower. In its juvenile, creeping form, ivy does not flower. If you like holly, grow a cultivar with berries – since holly has male and female plants, not all of them do, and remember that a female-sounding cultivar name does not guarantee a female plant. If you have an unknown holly that never has any berries, check the flowers to see which sex it is. If you grow apples, leave windfalls for migrant redwings and fieldfares – they are very hungry when they get here from Scandinavia, and they love *apples.*

Growing shrubs with smaller berries will provide for the smaller fruit-eaters. Elders are ideal, and both the red-berried Sambucus racemosa *and the black-berried* S. nigra *are available in a range of interesting cultivars with variegated, purple, gold and finely divided leaves. Remember, however, that bushes subject to the hard spring pruning recommended to produce the best foliage will not produce as much fruit. If you want lots of berries you will have to put up with your elders being a little untidy. Do not expect the smaller fruit-eating birds to be as obvious as the big thrushes – they are mostly more secretive in their fruit eating. Robins eat a lot of fruit, but it is not easy to catch them at it!*

Seed eaters such as finches and tits are best provided for by nuts, but they will eat seeds from flower heads too, so don't be too enthusiastic about cutting down dead stalks in autumn. Sunflowers and teasels are particularly favoured. And don't forget water – birds need to drink as well as eat.

Finally, for those Scrooge-like gardeners who don't want birds to eat their berries, stick to species and cultivars with white or yellow berries.

CATS

We can't really leave garden wildlife without mentioning cats, so sensitive cat owners should look away now. In an attempt to find out what Britain's pet cats are up to, the

Mammal Society asked nearly 1,000 cat owners to record all the animals their pets killed during the months of April–August 1997. During this period, the cats dispatched over 14,000 mammals, birds, reptiles and amphibians, including over 4,000 mice and nearly 2,000 voles. The top bird victim was the sparrow, but all the common garden birds were killed in large numbers, while frogs and slow worms were the most frequently killed amphibians and reptiles.

How well cats were fed made no difference at all – well-fed cats killed just as many prey as hungry cats, although fat cats generally killed fewer. Male cats killed more than females, but only slightly, and young cats killed far more than older ones. If you hope that your cat will earn its keep by killing rats, then think again – very few rats were killed during the survey period. This is hardly surprising, since adult rats are probably more than a match for the average domestic moggie.

Scaling up to the whole of Britain, and adjusting for lower rates of activity during the winter, pet cats probably kill around 250 million animals a year. While this may seem to be an extremely large number, it is notoriously difficult to attach any exact significance to such figures. Many animals killed by cats may have been sick, or very young or old, and would perhaps have died from other causes anyway. Before you comfort yourself with such thoughts, however, consider first that the cats in the survey did not confine themselves to common species. They also killed rare and protected species, including dormice, water voles, bats, treecreepers, goldcrests and sand lizards. Second, there are six times more pet cats than all other terrestrial predators combined – thirty-eight times the number of foxes, for example. Third, the kills reported in the survey are only those we know about. We don't know how many animals the cats were too lazy or devious to bring home, nor do we know how many are killed by Britain's almost 1 million stray and feral cats.

Finally, the birds actually killed by cats may not be the whole story. Ecologists are only just waking up to the huge changes in prey behaviour that may be caused by the mere presence of predators. Consider how your own shopping habits might change if you knew there was a finite chance of being killed and eaten every time you ventured out to the supermarket. Even if your cat never kills anything, there's increasing evidence that the fear caused by its presence may reduce the number of offspring raised by garden birds. Given that there are now 500 cats per square kilometre in Britain, about twice as many as thirty years ago, fear of cats must be added to the list of possible explanations for the otherwise puzzling decline in many urban birds such as house sparrows and starlings.

GARDENERS' TIP

If you own a cat and care about the wildlife in your garden (and your neighbours' gardens), there are two simple things you can do. Fit a bell or a CatAlert sonic collar, which reduces the numbers of birds killed by about half, and keep your cat in at night, which reduces the total kill rate by a remarkable 80 per cent.

If you don't own a cat and want to reduce cat visits to your garden, the RSPB recommend installing an ultrasonic cat deterrent in your garden.

Put bird feeders and nest boxes where cats cannot reach them or get close to them.

Acknowledgments

IN NO PARTICULAR ORDER, THANKS ARE OWED TO PHIL Grime, John Hodgson, Mick Uttley, John Hinton, Pete Smithers, Alison Loram, Kevin Gaston, Hans Cornelissen, Richard Smith, Phil Warren, Rosemary Booth, Stuart Band, Greg Masters, Jan Bakker, Renée Bekker, Begoña Peco, Costas Thanos, Marc Cohn, Steve Furness, Carla Bossard, Adel Jalili, Avice Hall, Sandra Díaz, Vladimir Onipchenko, Jill Rapson, Tony Hull, Martin Kent, Sue Hillier, Sarah Buckland, Mark Davis, Roberta Ceriani, Mark Hill, Peter Poschlod, James Hitchmough, Mary Leck, Richard Tapper, Andy Stevens, Liu Zhimin, Ann Conolly, Val Standen, Ric Colasanti, Jerry Tallowin, Nigel Dunnett and Chris Sydes. Special thanks go to Gaby Bartai Bevan, for giving me my first chance to write about science and gardening; to Tony Kendle at the Eden Project, who suggested I write this book and was then quite unfazed when the complete manuscript arrived the next day; to Mike Petty at Eden and Sally Gaminara at Transworld, who were kind enough to listen to Tony and generous enough to agree with his opinion; to Susanna Wadeson at Transworld for her enthusiastic support for this revised and expanded edition of this book. Finally, to Pat, Lewis and Rowan for their support (or at least tolerance) during the writing of this book, and their largely successful attempts to appear unsurprised that someone was willing to publish it.

Further Reading

Michael Proctor, Peter Yeo, Andrew Lack: *The Natural History of Pollination* (HarperCollins, 1996).
Max Walters: *Wild and Garden Plants* (HarperCollins, 1993).
B. N. K. Davis, N. Walker, D. F. Ball, A. H. Fitter: *The Soil* (HarperCollins, 1992).
D. Gledhill: *The Names of Plants* (Cambridge University Press, 1989).
Stefan Buczacki: *Understanding Your Garden: The Science and Practice of Successful Gardening* (Cambridge University Press, 1990).
Barbara and David Snow: *Birds and Berries* (Poyser, 1988).
Jennifer Owen: *The Ecology of a Garden: The First Fifteen Years* (Cambridge University Press, 1991).

Index